GOAT HUSBANDRY

GOAT HUSBANDRY

by DAVID MACKENZIE

Revised and Edited by
JEAN LAING

faber and faber
LONDON · BOSTON

First published in 1957
by Faber and Faber Limited
3 Queen Square London WC1N 3AU
Second edition 1967
Third edition 1970
Reprinted 1972, 1974, 1975, 1976 and 1978
Fourth edition 1980
Reprinted 1985
Phototypeset by Western Printing Services Ltd, Bristol
Printed and bound in Great Britain by
Richard Clay (The Chaucer Press) Ltd, Bungay, Suffolk
All rights reserved

British Library Cataloguing in Publication Data

Mackenzie, David, b. 1916
Goat husbandry. – 4th ed.
1. Goats
I. Title   II. Laing, Jean
636.3'908   SF383

ISBN 0–571–18024–8
ISBN 0-571-11322-2 Pbk

Contents

Illustrations

Plates

Figures

Tables

Preface to the First Edition

Certain references in this preface are to material omitted from later editions

Nearly twenty years ago, I retired into a converted hen house with a milking pail, a book of instructions, and an elderly goat of strong character. There was milk, among other things, in the pail when the goat and I emerged at last, with mutual respect planted in both our hearts. The book of instructions was an irrecoverable casualty.

No book is a substitute for practical experience, but books become more important as animals become more productive; the modern farmer and smallholder has to remember so much, so often and so quickly that each class of farm livestock requires to be accompanied by a comprehensive work of reference as a passport to the modern farmyard.

Holmes Pegler's *The Book of the Goat* was published first in 1875, and is at present in its ninth edition (now out of print); *Modern Milk Goats*, by Irmagarde Richards, was published during the first quarter of the present century in the U.S.A. In the meantime there have been published on both sides of the Atlantic a number of excellent practical handbooks on goatkeeping, goat farming and goat fancying, but no work with so wide a scope as *The Book of the Goat*. The omission is one of several handicaps to an expansion of commercial goatkeeping and always struck me as surprising until I found myself involved in the attempt to fill it.

The fact is that the study of the goat has commanded rather less scientific attention than the study of mice and guinea-pigs. The practice of goatkeeping in Britain, though successful in establishing world milk-production records and a vigorous export trade, is so highly variable, individual and empirical that it is almost indescribable. In search of information from scientific and technical institutions, the writer on goats is liable to be treated with the respect accorded to all things pertaining to the goat in Britain. While the majority of the leading exponents of goatkeeping practice in this country are eager to help, few have the time or temperament to measure and record their own practice with useful accuracy.

At several stages in the writing of this book I should have quailed before the obstacles so presented, but for the encouragement of Lawrence D. Hills and his fertile suggestions.

To Professor S. A. Asdell of Cornell University, U.S.A., and to Professor C. W. Turner of Columbia University, U.S.A., I am particularly grateful, not only for guiding me to the scientific work which laid the basis of the study of goat nutrition, but for their kindness to a stranger.

Many of my readers may agree with me in regarding the section dealing with problems of mineral imbalance in the goat as of greater practical use to the breeder than any other in this book; they will wish to join me in thanking Dr E. C. Owen of the Hannah Dairy Research Institute, and Brynmore Thomas of the Durham University School of Agriculture, who have provided me with much of the material on which this section is founded.

In my chapter on 'Goats' Milk in Human Nutrition', as a stockman and farmer, I found myself on rather unfamiliar ground; I am particularly grateful to Dr J. B. Tracey for reading my chapter and enabling me to enrich it with the fruits of his unique joint experience of goats and medical practice. For guidance in the special literature on this subject, I am indebted to Dr Mavis Gunther, M.A., M.D., of University College Hospital, London, whose kindness in reading and criticizing the chapter has greatly strengthened it. I was also assisted here by the advice of Sister E. Morrison, of *Nursery World*, based on her incomparable experience of the feeding troubles of 'difficult' infants, and by the classic work and personal assistance of F. Knowles, Honorary Analyst to the British Goat Society.

I would also express my thanks to many busy men of science who found time to give me help on technical points; to Stephen Williams, M.Sc., farms manager to Boots Pure Drug Company, for sources of information on the food capacity of the cow; to the Veterinary Advisory Department of Boots Pure Drug Company for information on the relationship between parasitic worm infestations and Vitamin B deficiencies; to James A. Paterson of the Scottish Milk Records Association; to Dr Fraser Darling; to C. A. Cameron, managing director of Alginate Industries (Scotland); to G. Kenneth Whitehead, Withnell Fold; to the Mond Nickel Company for information on cobalt, and to the Iodine Educational Bureau for a good deal of material on breeding troubles.

The short survey of industrial waste products suitable for goat feeding obliged me to trespass on the kindness of the technical executives of a number of firms; I thank Dr V. L. S. Charley of H. W. Carter & Company for a particularly happy correspondence on the use of blackcurrant pomace, and for generous assistance in investigating the matter; R. B. Norman of H. J. Heinz Company, W. E. Rhodes of Chivers & Sons, R. S. Potter of William Evans & Company, and R. E. Harris of Calindus Food Products for their helpful information.

I have particularly enjoyed the correspondence with friends of the goat in other countries, and hope that the list of names and addresses in Appendix I will not only aid the circulation of knowledge on goatkeeping but enable travelling goatkeepers to make contacts over a considerable proportion of the world. In compiling this list and for information on world goatkeeping I am indebted to Tanio Saito of Uwajima, Japan, S. de Jogn Szn., Secretary of the Goat Society of the Netherlands, Sigurd Andersen, Editor of the *Journal* of the Danish Goat Society, Mr Robert W. Soens, Secretary of the American Milk Goat Record Association, Monsieur C. Thibault, Director of the Station for Research in Animal Physiology, Jouy-en-Josas, France, and Mrs du Preez of Cape Province, South Africa and Dr Finci of the Department of Agriculture of Israel.

Of goatkeepers in this country I must first express my gratitude to Miss M. F. Rigg, the ever-helpful Secretary of the British Goat Society, whose assistance has been spread over the several years during which the book was being prepared, and has been indispensable. I acknowledge with gratitude the permission of the Committee of the British Goat Society to make use of photographs and many extracts from articles first published in the British Goat Society year books and monthly journals. I thank the many contributors to these richest sources of goat knowledge.

Of individual goat breeders I owe a special debt to Mrs J. Oldacre, Hanchurch Yews, for the photographs of yarded goats, the plans of a model goat-yarding system and goat garden prepared by her architect husband, and for some of the most carefully recorded information on practical goatkeeping that it has been my pleasure to use; J. R. Egerton for many beautiful photographs and for information on RM5 Malpas Melba; to Lady Redesdale for her cheese recipe; to Mrs Jean Laing, Moorhead, Newton Stewart, for photographs, for information on her feeding methods and for providing a shining example of economic methods of management; to Miss Mostyn Owen, Mrs Margaret Train and Mrs J. D. Laird for photographs; and to Miss Jill Salmon, Cothlan Barn, for some fine photographs and for that rare and precious material—accurately recorded information on feeding for high yields.

I am grateful to the Ministry of Agriculture and Fisheries and the Department of Agriculture for Scotland for the trouble they have taken to prepare the statistics of goat population in Appendix 2, and acknowledge with thanks the permission of the Director of Publications at Her Majesty's Stationery Office for permission to use material from 'Rations for Farm Livestock, Bulletin 48', and a design from 'Farm Gates' (Fixed Equipment on the Farm, Leaflet 8), as the basis of the goat-proof latch, illustrated in Fig. 10.

The photograph of the Wolseley Electric Fence, and the drawings of the Gascoigne Goat-Milking Machine, are kindly contributed by the firms concerned.

The reproduction of Bewick's Goat is by permission of the Victoria and Albert Museum.

The remaining drawings are by Kenneth Hatts of Bournemouth Art College. I feel more grateful to him each time I look at them.

The main brunt of book writing is borne by the author's household; I think particularly of my small son who, on the day after the manuscript was posted to the publishers, systematically destroyed the remaining stocks of stationery in the house to prevent me writing another.

Glen Mhor, Kishorn, Wester Ross   DAVID MACKENZIE
8 April 1956

Preface to the Second Edition

Certain references in this Preface are to material omitted from later editions

To the many goat breeders from many countries, who have sent me news, comments, problems and criticisms, are due my thanks and the credit for most of the improvements to be found in the new edition.

I am particularly grateful to Miss M. F. Rigg, Secretary of the British Goat Society, for helping me in various ways, and for assistance in investigating the circumstances and the extent of the serious decline in the productivity of goats in Britain.

To have had the help of Dr M. H. French, Chief of the Animal Production Branch of the Food and Agriculture Organization of United Nations, is a privilege which has enabled me to give an up-to-date evaluation of the importance of goats in the agriculture of developing countries, and to shatter the image of the Destructive Mediterranean Goat, which lies at the heart of so much anti-goat prejudice.

The therapeutic use of goats' milk, especially in cases of allergy to cows' milk, accounts for an increasing proportion of goats'-milk production in Britain and U.S.A. During the past ten years there has been a volcanic eruption of scientific evidence about the extent and gravity of cows'-milk allergy. This alarming material is very relevant to the future of commercial goat breeding and dairying, and I have tried to present it in a form usable by a goat farmer and respectable by medical men. Dr L. Sutherland, M.B., Ch.B., D.P.H., D.T.M. & H., has done her best to guide my pen in the way of objective truth in this matter. I do not wish to saddle her with any responsibility for statements made in this book, but with my gratitude.

By calling my attention to the newly discovered method of deodorizing male goats and eliminating the occasional 'goaty' flavour from goats' milk, Miss Jill Salmon has increased the load of gratitude I already owe her, and provided the answer to many a goatkeeper's prayer.

I confess that I lack the patience and devotion to listen for years on end to all the troubles, big or little, of every goatkeeper who cares to write or phone, and to search out a helpful and sympathetic answer to each one. The extent of my practical knowledge suffers as a result, but this book suffers relatively little, because I have been able to call on the prodigous volume of such practical knowledge that Mrs Jean Laing has accumulated. Like many goatkeepers in Scotland, I am thankful for it.

Finally a word of thanks to readers who are going to write to tell me what I have done wrong this time, and to ask me questions to which I do not know the answer. They keep the book alive.

Bridge of Urr, Kirkcudbrightshire   DAVID MACKENZIE
30 November 1965

Preface to the Fourth Edition

When I agreed, with some reluctance, to attempt the task of updating David Mackenzie's classic text on goatkeeping, which has become known to goatkeepers throughout the world as their 'bible', I had little idea what I was letting myself in for. Without a generous response from those who so kindly made time to help me, the task would have been impossible.

Where additions and amendments have been necessary, I have followed as nearly as possible the beliefs of the author. This has meant that I have in some cases seemed to put words into the author's mouth, as it were, but this has proved to be the most satisfactory way of maintaining the flow of the text without editorial insertions. Where I have omitted material included in previous editions this has, in the main, been on account of recent research and developments, and in order to present goatkeeping in the light of today's conditions.

Whilst appreciating the enormous contribution of the dedicated master breeders, whose skill has given us the high quality, sound conformation and long lactation of the dairy goat in Britain, and on whom we so much depend for our stud goats, David Mackenzie was primarily concerned with the goat as an economic household provider. He was also much concerned with the importance of goats' milk to those suffering from certain allergic conditions. His chapter on the nutritive and curative properties of goats' milk was thoroughly revised by him for the second edition, shortly before his death, and as it seems certain that he would have wished to revise it further for publication today, this chapter has been omitted. I have, therefore, at the publisher's request, added a short Appendix on 'Allergies and Goats' Milk'. I have added too an Appendix containing a few 'Notes for Novices' where advice on some of the problems commonly encountered is gathered together for easy reference.

My thanks are due to the British Goat Society for allowing reproduction of material from their Year Books and Journals, and to Bob Martin, the B.G.S. Shows Secretary, for his Appendix giving details of the revised regulations for Milking Competitions; to those who supplied new photographs; to Philippa Awdry, Pamela Grisedale, Bryce Reynard and Robert Haslam, for accounts of their goat-farming systems; to Mrs Leueen Hill of Redruth for her recipes for smallholder cheese and lactic acid curds; to Malcolm Castle and Paul Watkins for the information in Appendix 4; to Peter Wray, M.R.C.V.S., for his advice on disease and accidents; to Patricia Sawyer for taking the trouble to write out all those tricky names and addresses in various countries, and for her notes on export; and finally to Ann Rusby, who struggled so valiantly on her typewriter at home, and to Barbara Ellis for her painstaking editorial help, both of whom managed somehow to decipher my handwriting.

January 1980   JEAN LAING

Publishers' Note

Jean Laing might have wished to make minor revision to this new impression of the 4th Edition but, sadly, she died a few months before the reprint was put in hand. (In the circumstances, all that has been possible is some updating of the information given in Appendix Seven.)

The publishers take this opportunity of recording their appreciation of Mrs Laing's help and advice over a period of many years in connection with the reprinting of *Goat Husbandry*, and in particular for her invaluable revising and editing of the 4th Edition.

September 1985

Introduction (1956, revised 1965)

When man began his farming operations in the dawn of history, the goat was the kingpin of the pastoral life, making possible the conquest of desert and mountain and the occupation of the fertile land that lay beyond. The first of man's domestic animals to colonize the wilderness, the goat is the last to abandon the deserts that man leaves behind him. For ever the friend of the pioneer and the last survivor, the goat was never well loved by arable farmers on fertile land. When agriculture produces crops that man, cow and sheep can consume with more profit, the goat retreats to the mountain tops and the wilderness, rejected and despised—hated too, as the emblem of anarchy.

During the last hundred years much has been done to improve the productivity of goats in Northern Europe, North America and Australia, where the modern dairy goat can convert the best of pastures and fodder crops into milk, as efficiently as the modern dairy cow. Like most small production units, the goat is expensive with labour, but in its use of raw materials it rivals the cow, even when the raw materials are those best suited to the cows' requirements.

In developing countries, where there is normally an embarrassing surplus of suitable labour, the high labour requirement of goat dairying is a social virtue. If the land be arid or mountainous, the goat may prove to be the only economic source of milk. The authors of national development plans and international aid projects are not so starry-eyed today as they were ten years ago; the cow dairy farm in the desert and the steel mill in the jungle are giving way to reality and goat improvement schemes.

In the advanced countries, medical research has been discreetly lifting the lid off the consequences of our peculiar practice of snatching the newborn infant from its mother's breast and fostering it on the first available cow. The consequences may be grim, or may persist throughout life, unless goats' milk comes to the rescue. In Britain and the U.S.A. there is a growing demand for goats' milk for therapeutic use, which cannot afford to be deterred by the higher labour costs of goats' milk production.

Unfortunately, this new resurgence of interest in goat dairying may, in the present state of goat breeding, wreak havoc. The last great resurgence of goatkeeping met the wartime challenge of food shortage, and culminated, in 1949, in the legal black market in dairy produce. From that date we mark a steady decline in the quality of the yields of pedigree goats in Britain, a decline which still continues. The new goatkeepers of 1949 were taught to feed their goats as miniature cows, and proceeded to breed them selectively for their response to this feeding. With honourable exceptions, the goat breeders of 1965 maintain this destructive tradition. An expansion of goat breeding within this tradition can only result in establishing a strain of goats which do, in fact, perform like miniature cows. A good cow, miniaturized to goat size, could produce no more than 6 pints a day on the best feeding. A good goat can produce twice as much.

The relationship between size and efficiency in all productive farm animals is so well established in both theory and practice that, when confronted with the performance of the modern dairy goat, the nutritional scientist and the farmer tend to regard it as a somewhat indecent Act of God, unrelated to His regular arrangements. For the rule is adamant: provided the feeding is sufficient, the big animal must outyield the little one; the big one has a smaller surface area in proportion to its bulk and potential food capacity, and so uses less of its food to keep itself warm and more to make meat or milk. Friesians replace Ayrshires as pastures are improved; low-ground sheep are bigger than mountain breeds; every beast, ideally, is as big as its pasture permits. But fifteen 1-cwt [50·8 kg] goats will make rather more milk out of the ration of a 15-cwt [762 kg] Friesian cow than the cow can. Yet the rule is unbroken; for it applies only between animals of the same species: a kind Providence has decreed that goats are very far from being miniature cows.

A goat, however 'modern' and 'dairy-bred', is a goat, a member of the species familiarized in nursery picture books and biblical illustrations, target of laughter and abuse for countless centuries, Crusoe's salvation and mankind's first foster mother, the Common Goat.

The processes of history have greatly reduced the goat in Britain; agricultural textbooks have exiled the hardy ruffians for half a century; scientists have used the modern dairy goat as an expendable model cow, but done little to investigate the basic attributes of the goat as such. The purpose of this book is to drag this half-mythical creature out into the light of present-day animal husbandry, that we may know it, use it and care for it more effectively.

We must begin by evicting from our minds the false analogies between goat and cow and between goat and sheep. We can hardly do better than refer back to Thomas Bewick's *History of Quadrupeds*, published at the beginning of the nineteenth century, when the Common Goat was still common in Britain.

'This lively, playful and capricious creature occupies the next place in the great scale of nature; and though inferior to the sheep in value, in various instances bears a strong affinity to that useful animal.

'The Goat is much more hardy than the sheep, and is in every respect more fitted to a life of liberty. It is not easily confined to a flock, but chooses its own pasture, straying wherever its appetite or inclination leads. It chiefly delights in wild and mountainous regions, climbing the loftiest rocks and standing secure on the verge of inaccessible and dangerous precipices; although, as Ray observes, one would hardly expect that their feet were adapted to such perilous achievements; yet, upon nearer inspection we find that Nature has provided them with hoofs well calculated for the purpose of climbing; they are hollow underneath, with sharp edges like the inside of a spoon, which prevent them from sliding off the rocky eminences they frequent.

'The Goat is an animal easily sustained, and is therefore chiefly the property of those who inhabit wild and uncultivated regions, where it finds an ample supply of food from the spontaneous production of nature, in situations inaccessible to other quadrupeds. It delights in the healthy mountain or the shrubby rock, rather than the fields cultivated by human industry. Its favourite food is the tops of the boughs or the tender bark of young trees. It bears a warm climate better than the sheep, and frequently sleeps exposed to the hottest rays of the sun.

'The milk of the Goats is sweet, nourishing and medicinal, and is found highly beneficial in consumptive cases; it is not so apt to curdle on the stomach as that of the Cow. From the shrubs and heath on which it feeds, the milk of the Goat acquires a flavour and wildness of taste very different from that of the Sheep or Cow, and is highly pleasing to such as have accustomed themselves to its use; it is made into whey for those whose digestion is too weak to bear it in its primitive state. Several places in the North of England and in the mountainous parts of Scotland are much resorted to for the purpose of drinking the milk of the Goat; and its effects have been often salutary in vitiated and debilitated habits.

'In many parts of Ireland and in the Highlands of Scotland, their Goats make the chief possessions of the inhabitants; and in most of the mountainous parts of Europe supply the natives with many of the necessaries of life: they lie upon beds made of their skins, which are soft, clean and wholesome; they live upon their milk and oat bread; they convert part of it into butter and some into cheese. The flesh of the Kid is considered as a great delicacy; and when properly prepared is esteemed by some as little inferior to venison.

'The Goat produces generally two young at a time, sometimes three, rarely four; in warmer climates it is more prolific and produces four or five at once. The male is capable of propagating at one year old and the female at seven months; but the fruits of a generation so premature are generally weak and defective; their best time is at the age of two years or eighteen months at least.

'The Goat is a short lived animal, full of ardour, but soon enervated. His appetitie for the female is excessive, so that one buck is sufficient for one hundred and fifty females.'

Thomas Bewick's account of the goat suffers little from the passage of nearly 200 years. The wildness of taste which he attributes to goats' milk can be tamed by dairy hygiene, mineral supplement and surgical operation, but many newcomers to goats' milk can still catch his meaning. The resident population of the 'wild and uncultivated regions' has been eroded by hunger and administration, but the 'heathy mountain and shrubby rock' are still good dairy pastures for a goat. Bewick's few paragraphs contain clues to the peculiarities of goat digestion, housing requirements, and control, and to the phenomenal productivity of the modern dairy goat.

In following up these clues in subsequent chapters, the assumption is that goat farming can and will develop into a considerable branch of agriculture. As such, goat farms must be mainly in the hands of established farmers with a general knowledge of crop and stock, and utter ignorance of goats. Such knowledge and ignorance is assumed; but a chapter on cropping for goats is included to help the domestic and small-scale goatkeeper in whose hands the goat tends to be most profitable.

Chapter One

The Place of the Goat in World Agriculture

Cow, sheep and goat, all provide man with meat and milk. At times we are inclined to think and act as though they were rivals for that dubious honour. In fact they are prehistoric grazing companions, who need each other's help to make the most of available pastures.

The natural covering of the earth ranges down from the unbroken canopy of high forest to the small, ephemeral herbage of the desert, hot or cold. The permanent natural grasslands lie next to the desert fringe: steppe, prairie, pampas, mountain top and sand dune grow only grass because the soil is too thin, dry, cold, to grow anything bigger. Their vast extent supports large herds, but the typical stocking capacity of such pastures is only a cow and calf to thirty acres; under these conditions sheep, cow and goat are as competitive as they are complementary and the productivity of the land is inevitably low.

The best natural pastures are temporary, being stolen from woodland by fire, drought, flood and storm. There is good natural grazing too on savannah-type land, where soil and climate maintain a precarious balance between 'bush' and grasses. Many pastures are man-made; the best of them, in New Zealand, Holland, the English Midlands, being derived from marsh or woodland. Temporary grazings, whether natural or man-made, can be maintained only by the co-ordinated efforts of grazing stock and by man.

The reversion of such temporary pastures to scrub, woodland and forest is pioneered by coarse grasses and unappetizing weeds, rushes, thistles, brackens, nettles, etc.; these are followed by still more repellent small shrubs, bramble, briar, gorse and thorn. Within the bridgehead so established, the windblown seed of light-leaved trees can germinate and grow, to provide a refuge for beasts and birds that carry the seed of the forest giants, in whose shade all rivals perish.

The first line of pasture defence is the sheep, whose split lip enables it to bite herbage down to soil level. The sheep catches the toughest of invaders in their seedling stage, while they are still tender and nutritious. Its daily capacity for food intake is smaller in relation to its size than of its companions, so its grazing habits are more selective; the sheep generally avoids coarser vegetation, but exercises some control of established shrubs by nipping out the soft growing-points.

The cow crops the pasture evenly and systematically, leaving behind it a sward which the sheep can clean to the bone. In relation to its size the cow's food capacity is only slightly greater than the sheep's but, having the economic advantages of a larger unit, it can afford to accommodate coarser fodder. However, the sweep of its tongue must embrace $1\frac{1}{2}$ cwt (76·2 kg) of grass each day, and the cow has neither time nor patience for anything that frustrates the steady rhythm of its grazing, be it short herbage, prickly weeds or woody shrubs. When the invading seeds get past the seedling stage the main defence against reversion is the goat, assisted in recent centuries by man.

The goat faces its task with a hero's equipment. It has the toughest mouth of all the ruminants and can consume with profit and pleasure such well-protected vegetable treasures as the bramble, briar, thistle and nettle. In proportion to its size the goat can eat more than twice as much fodder each day than either the sheep or cow, almost one-third of its total body capacity being available to accommodate food in the process of digestion. Because its fodder accounts for so much of its weight, the goat's need of actual nutrients to maintain its own condition is slightly less than that needed by a sheep of the same size. So the goat can consume a large quantity of coarse fodder wherein the actual nutrients are very dilute; or, if it has access to fodder of good quality, the goat may find itself in possession of such a large amount of nutrients surplus to its maintenance requirements that when suitably bred it can be prodigiously productive of milk and offspring.

Fig. 1 illustrates the relative grazing capacities and needs of the three species. The lightly shaded areas indicate the relative daily food capacity. The dark squares inside the shaded area show the relative amounts of nutrients needed to maintain the animal's condition, without any allowance for growth or production. Grass capacity varies within the species; for cattle it varies between a normal 12·5 per cent to an exceptional 20 per cent of bodyweight; for sheep between 12·5 per cent and 15 per cent; for goats between 25 per cent and 40 per cent normally and up to 50 or 55 per cent exceptionally. The cow is drawn to a scale one-tenth of that used for both sheep and goat; had I drawn a half-ton cow and a half a ton of goats and sheep on the same scale, the goats would still be seen to have eaten twice as much as the cow or sheep.

Even in a temperate climate such as that of France, and on cultivated pastures, the goat has been shown to use 15 per cent more

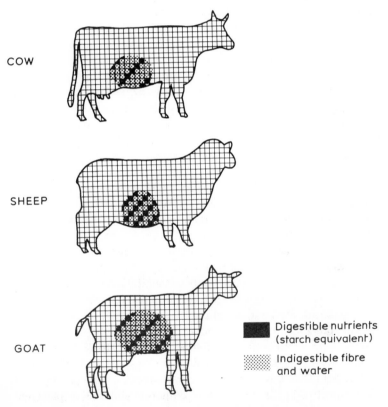

COW

SHEEP

GOAT

Digestible nutrients (starch equivalent)

Indigestible fibre and water

Fig. 1. Grazing capacity and minimal fodder quality requirement

varieties of available pasture plants than either sheep or cattle. Vegetation which is exposed to great heat, drought or frost must protect itself from freezing or evaporation with a tough fibrous skin or texture. If sheep or cattle are to graze such pasture at all, they will benefit from the presence of goats to control its more fibrous elements. When conditions become extreme, the goat is left in sole possession.

The wild ancestors of the domesticated goat are *Capra aegagrus* of Persia and Asia Minor, *Capra falconeri* of the Himalayas, and *Capra prisca* of the Mediterranean basin. The common goat of most of Europe and Asia is derived from *Capra aegagrus;* the Kashmir and Cheghu goats from *Capra falconeri*; the Angora goat from a cross *aegagrus* and *falconeri*. The remains of the domesticated offspring of *Capra aegagrus* have been found in Early Stone Age deposits in Switzerland, and it is to be presumed that this is the only type to have reached northern Europe until modern times. But investigations in Egypt by Professor de Pia have shown a prehistoric dwarf goat replaced by domesticated derivations of the twisted-horned *Capra prisca* before the advent of modern derivatives of *Capra aegagrus*. It is probable that *Capra prisca* provides some of the ancestry of most modern goat stocks in the Mediterranean basin, and that this accounts for some of their distinctive characteristics.

The goat is valued mainly for its meat and milk. As a milk producer the goat is inevitably more efficient where the available fodder is of such low quality that a cow can barely live. On the desert fringes of the Middle East the cow doesn't get a look in; milk supply is in the care of the Mamber goat and its relatives. The Mamber is a large goat, weighing up to 120 lb (54·4 kg) with long black hair which is used mixed with wool for carpet making; she yields up to 5 pints (2·8 litres) a day when well fed.

The cow dairy business does not start until the quality of fodder is such that the cow can give 2 gallons (9·1 litres) of milk a day at the peak— that is, feed her calf and provide a gallon (4·5 litres) a day for sale. On a lower level of feeding than this, a goat can feed a kid and provide about 3 pints (1·7 litres) of milk at the peak to her owner (i.e. where starch equivalent of feeding is 32 to 36 per cent of dry matter). The advantage of the goat is further extended by her fecundity: two kids are normal in temperate climates; triplets and quads are common with well-fed goats in warm climates; one or more of these provides a valuable carcase at a fortnight old. But it is significant that in the countries where they are most extensively bred, the average yield of goats is just at this marginal line of about 3 pints at the peak, while suckling kids under control. Such is the figure given for the two main breeds of Indian goats, the Jumna Pari and the Beetal, though individuals of both breeds have proved capable of yields of 6 (3·3 litres) and 7 pints (3·9 litres) a day and over 100 gallons (454 litres) a year. Where lower yields are prevalent, as with the dappled Bar Bari goat of Delhi and the little black goat of Bengal and Assam, the Breeds are dual-purpose, with a smaller, meatier body and higher fecundity.

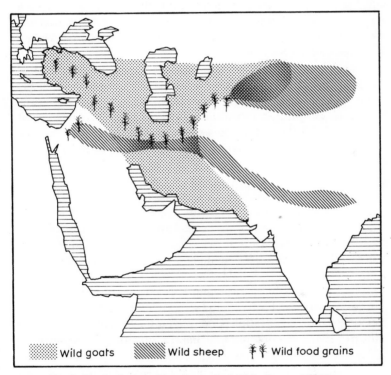

Fig. 2. The beginning of farming in the Middle East *c*. 10,000 BC

Where yields rise above the 3-pint level, indicating a standard of living that would support a dairy cow, as they do with the Zariby goat of Egypt in some districts and with Granada and Murcian goats of Spain, the cow may be kept out of business by the difficulty in distributing its milk on bad roads in a hot climate. The goat delivers on the hoof, in household quantities; the cow is essentially a wholesaler.

The goats of Malta are to a great extent protected from the competition of dairy cattle. Their yields average 3 pints; but the 3 pints is produced by a 70 lb (31·7 kg) body, and the occasional full-sized goat has proved capable of yields of over 300 gallons (1,364 litres) a year.

It is when we come to the mountains where the pasture would do justice to a pedigree Ayrshire, but is at an angle at which no cow can graze, that we find substantial numbers of high-yielding milk goats. The Swiss Saanens, Toggenburgs and Chamoisées, also their counterparts on the eastern side of the Alps and the 'Telemark' goats of Norway, are all capable of yields of up to a gallon a day with a lactation of 8 to 10 months.

From these mountain breeds derive the substantial number of scattered herds of high-yielding goats which are found about the industrial cities of northern Europe and in isolated rural communities in fertile agricultural districts. These herds owe their existence to the fact that, given the inherited capacity of milk production, the goat is a slightly more efficient converter of pasture into milk than the cow.

Preposterous at first sight, this claim has a clear enough theoretical basis, and has been exactly demonstrated in practice by the experimental work sponsored by the Department of Agriculture of the U.S.A. The claim is fully dealt with in a subsequent chapter (p. 106); here it is sufficient to explain how it comes to be true.

From the point of view of comparative costing, the maintenance ration of the cow or the goat is the amount of food each requires to keep her going, while converting a given quantity of digestible nutrients into milk. Each 100 lb (45·4 kg) of goat requires 1½ times as much maintenance ration per day as each 100 lb of cow. But each 100 lb of goat eats twice, or three times, as much digestible nutrients a day as each 100 lb of cow. Consequently each 100 lb of goat has more digestible nutrients available for conversion into milk each day than each 100 lb of cow. For every 10-stone (63·5 kg) bag of dairy cake fed to it, the average goat produces 2 gallons (9.1 litres) more milk than the average cow.

But the goats' milk cannot easily be produced in commercial quantities to compete in cost with that produced by cows on good land. A well-managed herd of goats will produce a yield of 200 gallons (909 litres) per head; management and cropping for thirty of them is as laborious as the work of a dairy farm with twenty cows, yielding 600 gallons (2,727 litres) a head. So the labour cost per gallon goats' milk is roughly twice the labour cost per gallon of cows' milk. If the goats are sustained largely on natural waste vegetation, instead of crops, labour costs fall, feeding costs fall, and competition is possible. The man who doesn't charge himself anything for his own labour can usually produce goats' milk considerably more cheaply than anyone can sell or produce cow's milk. Moreover, while few Durham miners could buy or graze a cow, in the days when they were grossly underpaid, many of them were keen goat-keepers. The same principle holds good the world over; goats exist to cut the cost of living for many millions of hard-pressed families in cow-dairying country.

In France and in Norway special value is attached to goats'-milk cheese and in the United States a specialized market for goats' milk has been developed based upon its higher digestibility and its value to sufferers from allergy to cows' milk.

There is a continual fluctuation in goat populations throughout the world. Prior to 1936 Greece had almost three times as many goats as sheep; the numbers dropped 75 per cent in fourteen years; most of their former pastures were designated for afforestation in 1936, and goat grazing therein forbidden by law; such goats as survived this enactment helped substantially to feed the wartime partisans. In 1942 Thailand was cut off from supplies of condensed milk by the stringencies of war, and large-scale goat dairying rapidly grew up around Bangkok; the enterprise survived the advent of peace.

Apart from such dramatic movements in goat population, there is a steady response to world-wide and local fluctuations in the relative value of labour and food. Improved agricultural methods lower the price of food, and by increasing the value of land increase the labour involved in controlling goats; at the same time higher levels of land fertility can support the rivals of the goat in commercial dairying. While the goat is common in the Early Stone Age deposits of the Swiss lake settlements, it becomes rarer in the later ones; throughout history large-scale goatkeeping retreats before the advance of agriculture. On the other hand, scarcity of food forces interior land into production and demands the utmost use of available fodder; so we find a sharp rise in the goats' representation among farm stock in all countries during war and economic difficulty.

Broadly speaking, therefore, the goat's place in the world dairy business is primarily on land that is too poor, too hot or too steep to support dairy cattle; where such areas are extensive, a daily peak yield of 3 to 4 pints (1·7 to 2·3 litres) per goat provides a basis for commercial production, and in most cases is the maximum that the pasture will sustain. On better pastures, goats of good milking strain are capable of converting fodder to milk quite as efficiently as the best strains of dairy cattle; but as the labour cost per gallon of goats' milk is twice that for cows, this is a solid barrier to commercial goats'-milk production, which can be overcome only when goats' milk commands a higher price than cows' on the strength of its special merits or because of overall milk shortage, or where otherwise valueless and uncultivated land supplies the goats' feeding. Where labour is costless, as it often is when the goat is used for domestic supply, goats' milk is far cheaper than cows'.

In considering the goat as a meat producer, we must differentiate between kid meat and goat meat. Kid meat is a by-product of the dairy goat, which is probably appreciated and fully utilized in every country in which goats are kept except Great Britain and the 'White' Commonwealth countries. Intrinsically it is in every way as valuable as veal, and is rather more versatile in the hands of a good chef. Goat meat, when reasonably well produced, is in no way inferior to the general run of old ewe mutton, which was parsimoniously inflicted on the British dinner plate for fifteen years of war and 'austerity': but it cannot rise to the heights of prime beef or lamb and when it is produced in quantity it is produced for its own sake.

As beef cattle can rear a calf and acquire some degree of fatness on pastures too hot or poor to sustain dairy cattle, it follows that the breeding of goats for meat production is less extensive than the range of the dairy goat. In many areas where goats are the source of milk supply, beef and mutton dominate the menu. There remain regions of desert, dense jungle and high mountain pasturage where goats are both meat and milk to the inhabitants. Such is the case in Arabia, Syria, Iraq, in parts of Equatorial Africa and in the Himalayas. There are wider and more fertile regions where the value of cattle as draught beasts is such that it denies their use for meat. Throughout most of India the cow is sacred. Religion has endorsed the need to safeguard the producer of agricultural power units from the hunger of a chronically famine-stricken population. Goat takes the place of beef on the Indian menu; it is a happy arrangement: the firm, rather dry flesh of the goat takes well to currying, and is none the worse for its dryness in a land rich in relatively cheap vegetable oils. The wool of the sheep constitutes its main value in hot countries where its meat never reaches perfection; under tropical and sub-tropical conditions the goat is more palatable as well as more prolific.

On the other hand, in a cooler and wetter climate, the drier flesh of the goat is not so popular as the interlarded meat of its farmyard rivals; its fecundity is less prounounced and both sheep and cattle can better withstand a combination of cold and wet. Goat-meat production is, therefore, confined to hot countries and excessively high mountains; only the Angora goat, as a sideline to the production of mohair, supplies the tastiest of all goat meat in cooler regions.

The skin, wool and hair of the goat strengthen its economic basis in many areas throughout the world. Goatskin makes leather of the first quality; glacé kid shoe leather alone accounts for almost half the annual world consumption of skins. The long-haired goats of cooler climates produce the heavily grained skins which make up into the morocco leather used for bookbinding, furnishing, handbags and fancy leather work; second-grade skins are used for lining shoes. Angora skins dressed with the hair on are used for rugs and made up into fur coats. Altogether these miscellaneous uses take the skins of a considerable proportion of adult goats each year. Kid skins, which come mainly from southern and eastern Europe, are used principally for the manufacture of top-quality gloves. In 1977 the world production of goatskins totalled 329, 423 tonnes.

The woolly undercoat of the cashmere (Kashmir) goat, which is 'farmed' by nomadic tribesmen above the 15,000-foot contour of the eastern Himalayas, has a superlative quality. The annual wool crop is 3 to 4 oz (85 to 110 g) per goat and is combed out with scrupulous care, the process taking eight to ten days. The value of the animals to their owners probably lies mainly in their meat and milk. The Cheghu goat of similar type of habitat in the Western Himalayas is also used as a pack beast. In both cases the long outer coat is shorn and used for making tent-cloth, and mixed with other fibres for carpet-making.

Throughout the Middle East, Eastern Europe and India the hair of long-haired goats is used for weaving coarse cloths, making rugs and carpets and for packing mattresses. But the hair of the Angora goat, which originates from Turkey, is of a special quality, being a fine, lustrous silky fleece with an 8-in (20 cm) staple, which covers the animal to its hocks, and weighs rather more than an average Blackface sheep fleece. This material, the mohair of commerce, has properties that combine high lustre, good resceptivity to dyes and great durability. Plush and braid are less fashionable than they used to be and nylon has similar qualities, but mohair still meets a ready demand, providing the basis for furnishing velvets, for many of the better-quality fur fabrics, and for some popular flights of textile fancy in the form of ladies' scarves, besides making a big contribution to the carpet and rug industry. Mohair is the *raison d'être* of several million goats in South Africa, in the dry states of North America and in Turkey. The Angora goat thrives best in a fairly warm climate with a rainfall of under 20 in (51 cm) per annum.

From time to time, changes in economic conditions or agricultural techniques deprive the goat of its usefulness in an area where it has long played an important role. Swiftly and without fuss or publicity, the goat flocks dwindle and disappear. Recently this has been happening in Switzerland, where Alpine goat dairying can neither offer its labour force a wage, or way of life, to compete with the attractions of industrial employment; nor economize in labour by mechanizing its processes. So the goat flocks of the seven Swiss breeds are dwindling away at an accelerating rate Actually, this reduction has now stopped, and a stabilization can be expected. Two reasons explain this fact: first, there is a good demand and a high price is given, for goat cheese and kid meat; second, the goat is recognized as having an important role in the care of the landscape in the Alpine region.

On the other hand, when we encounter an organized propaganda campaign against goats, prominent officials demanding the extermination of goats, and laws directed against goatkeeping, we can be sure that goats are indeed an economic proposition for their owners in the country concerned; and that their owners lack political power. In the countries of the East Mediterranean and Middle East such a situation was common in the day of colonial rule, and persists today wherever government is in the hands of aliens or more prosperous classes. Throughout this area at the crossroads of the continents the land has been subjected to man's ill-treatment more intensively and persistently than anywhere else on earth. For thousands of years every patch of watered soil has been cropped and cropped again, without any manurial treatment, until finally abandoned, exhausted, unseeded and naked to the elements. Starting from the lowland plains, man has continued the process up the mountain side; but on the slope the naked soil of abandoned fields is quickly washed away in erosive torrents that gouge great scars in the plains below. The primitive techniques of shifting cultivation still persist. A 1961 Lebanon survey showed half an acre (0·2 ha) abandoned farmland for every acre in cultivation; in Greece, Turkey or Syria much the same was true; $1\frac{1}{4}$ acres (0·5 ha) of felled forest for each acre (0.4 ha) under timber, said the Lebanon survey. Where have all the cedars gone? Down the long road of international commerce that has raped this land without ever fertilizing it.

The pastures that once fed a mixed herd of cows, sheep and goats, were taken for crops, cropped to exhaustion and abandoned naked to the harsh mercy of the climate. The wretched cloak of scrub, spattered about the eroded land today, is all the fodder for the grazing herds. Only goats can scrape a living from it.

In Greece, however, where the goat was said to have been responsible for deforestation in some parts, notably Crete, the goat population fell drastically after the Second World War until, by the early 1970s, it was felt that the numbers were below what was desirable. Accordingly, the Greek government sponsored a programme of expansion which, with the removal of the ceiling price on meat—including that of goats—in the middle of the decade, helped to build up numbers again.

It is nonsense to suppose that trees are the only or the best counter to erosion. On Mediterranean hills the combination of hot sun, low rainfall and thin soil is unfavourable to tree growth; a turf of deep-rooting grasses fortified by a scattering of drought-resistant evergreen shrubs is the best protection against erosion that soil and climate permit, over wide areas, and is quite probably more effective than forest cover, if only because it regenerates so quickly after being burnt. As a British experiment in Tanganyika proved, in grazing containing bush and scrub the goat frees more grass than it eats. Grass on stock pastures carries the larvae of internal parasitic worms. Worm larvae are liable to desiccation, so they keep mainly to the layer of dense vegetation close to the ground; the higher branches of bushes and shrubs are relatively free from larvae; in self-defence, the goat prefers them. Some years ago, in Venezuela, a cattle-ranching lobby prevailed on a cattle-ranching president to decree the extermination of goats over a vast area. The law still stands, but the goat is now back to its former strength and the government is engaged on a goat improvement scheme. In the goat's absence, the bush invaded the pasture, grass was smothered, cattle stock deteriorated and ranchers' profits fell.

Let us grant that goats wreak havoc in a young plantation and prevent natural regeneration in a felled forest; men seeking firewood and raw materials for Mediterranean do-it-yourself furniture, have the same effect. Neither does much damage to mature timber. If trees must be planted in a place, goats must be excluded, and unauthorized men as well. But in the present context of human starvation and chronic malnutrition, prevalent throughout the East Mediterranean, there is a lot more need for goats than for trees. Goats' milk, goats' cheese and goats' meat are the main source of protein for the underfed and protein-starved majority in these countries. International aid, central planning, technical advice and political speeches are no substitute for protein in breaking the prevalent lethargy that baulks development there. Though goat owners are almost all poor men, goats are, on the average, 50 to 75 per cent more profitable than cows for meat and milk production under East Mediterranean conditions. Sheep are no substitute; their milk is rich in fat, but vegetable oil is cheap; in milk-protein production they cannot compete with goats; mutton is a luxury. The attempt to replace goats by sheep, cows or trees, in this area, is merely a rich man's racket allied to bureaucratic laziness. It is easier for the bureaucrat to blame the goat than to pin the responsibility on the real culprit, to ensure that cropped land is properly manured, and sown out to grass, instead of being abandoned to weeds when its cropping capacity is exhausted.

For many years to come the goat must retain its pre-eminence in this part of the world, and in others with a comparable climate. But the present goat population is excessive. By improving the qualiity of the stock, fewer goats could make better use of the available grazing. A considerable percentage of the flocks at present are surplus males or aged females, kept not for their productivity, which is nil, but to provide social security insurance against the risks of disease, famine, and marriage in the family. To convert this surplus into dependable currency and sausage would be a kindness to man, land and goat. Better organization of the marketing and distribution of goat products would do away with the unproductive section of the flocks, and improve the health of a protein-starved human community.

Once goats are recognized and treated as playing a vital part in the national development plan, it is possible and necessary to tackle the other objection to their existence in this part of the world—their liability to carry Malta fever and T.B. In fact goats, cows and pigs are all liable to infection with their own variety of brucella bacteria; all of them can pass the infection on to man. Man becomes partly immunized to the forms of brucellosis to which he is regularly exposed; but a world of travellers needs protection. Brucellosis can be eradicated from the Mediterranean goat as it is now being eradicated from the British cow. In the meantime, pasteurization of the milk of untested animals is a sure safeguard.

The goat's place in world agriculture is primarily in the 'developing' countries, and its future depends on the direction that development takes. This refutation of the standard slanders against the goat is based on the reports and policy discussions of the headquarters of the Food and Agriculture Organization of United Nations, the main channel of aid and advice to developing countries. For the first time for many centuries the goat has won some friends in high places. Who befriends the poor, befriends the goat. As humanity grows more humane, the goat grows in stature.

Chapter Two

The Goat in Great Britain

Until the late eighteenth century, goats were a normal source of milk supply for cottagers throughout the country, and featured on every common in England. In the moorland and mountainous districts of the North of England, Scotland and Wales goats occupied a key position in the rural community.

On the hills sheep were kept mainly for their wool, much of which went to clothe their owners; cattle provided the cash income when sold for slaughter in the autumn, and perhaps a summer milk surplus for conversion into butter and cheese; goats bore the main brunt of domestic meat and milk supply. In these districts today you may travel for hundreds of miles without seeing a goat, but can hardly travel more than a few hundred yards without passing a spot whose local place name testifies to its former popularity with goats. On a modern Ordnance Survey map of the Highlands of Scotland you will find more place names with the 'gour' and 'gobhar' theme (Ardgour, Arinagour) than you will find goats in all Scotland.

On this thriving goat population there descended the two-edged sword of industrial revolution and improved agriculture. Improved agriculture brought in its wake the enclosure movement; the extensive common grazings, whereon the peasantry of England's cattle, ponies, sheep and goats overcrowded inferior grazings, were put under the plough and crops. This immense development of Britain's food-producing capacity took at times the dramatic form of high-powered robbery by the local landlord, accompanied by local disturbances and migration from the land. But in the main it was a long, peaceful process, whereby the peasant exchanged a penurious independence for a wage through which he shared in the increased productivity of his lost commons. In either case it involved a large drop in the goat population. Where the goat survived in the neighbourhood of enclosed land, it was a nuisance to the farmer of the fields it raided; and a nuisance to its owner, who was obliged to control it by constant herding or tethering. As the value of agriculture labour increased with the fertility of the fields, goatkeeping became more expensive; the practice of tethering inevitably increased the goat's liability to infestation with internal parasites, and reduced its productivity. We may be sure that then, as now, the most enthusiastic goatkeepers were the most awkward individualists in the area, and their goats the *casus belli* with their farmer neighbours. Where the enclosure movement took on its more dramatic forms, the goats were the spearhead of the underground resistance movement of the cottagers.

The face of agricultural advance was turned against the goat; and the goat chivvied the heels of advancing agriculture with its infuriating impudence and eternal disdain. To the undying credit of its nuisance value, an atavistic hatred of goat still lingers in the mind of England's farmers. Hated by the masters of the land, it declined steadily in the affections of the land's servants to the point of near-extinction.

Few tears need be spent on the departure of the goat from the common grazings of English lowlands. For the most part it was replaced by sheep and cattle that could produce milk and meat and wool more economically, and by crops that greatly increased England's stocking capacity for man and beast. The change was the first necessary step towards industrialization. Certainly neglect of the goat was carried too far; there was scarcely any development of breed type or methods of management for one hundred years of changing agriculture; much land that goats could use was left useless, and many an isolated cottager, whom goats' milk could have sustained, went milkless. But as the prime role of British agriculture became the feeding of the urban populations, the goat lost for ever its place on England's better farming land.

The decline of goatkeeping in the mountainous and moorland areas was and is something of a national disaster. For the widely distributed inhabitants of these regions, the goat remains the most economical form of milk supply; the better use it makes of coarse fodder and its longer lactation have never failed to give it an advantage over contemporary cattle stocks. The reasons that the goats left the hills are therefore of more interest. They are perhaps best seen by a glance at the goats of the Scottish Highlands, where conditions are extreme and the extinction of the goat most dramatic.

Towards the end of the eighteenth century the Scottish Highlands supported approximately ten times the population that they do today; the basic mode of living was a very low standard of subsistence agriculture, with crops of primitive bearded oats and 'bere' barley and a stock of cattle, ponies, goats and a few sheep. Cattle were bred primarily for beef and provided the only substantial cash crop and export of the area. The sheep were the old tan-faced breed, of little value as mutton makers, but producing wool of the modern Shetland type for domestic consumption. The cattle provided a little surplus milk during the summer, and barren cows and bullocks were also occasionally bled at this season. Otherwise domestic needs of meat and milk were met by the goats. From May to September the whole stock was driven away from the arable land about the croft houses, and herded from summer sheilings on the high hills by the young folk of the community. The cheese and butter made at the sheilings was the mainstay of the people's winter rations.

Improved agricultural methods began to filter into the Highland area from about 1750, and their most successful application took the form of breeding the improved Border sheep on the Highland hills. Looked at from a national point of view, this enterprise provided a meat surplus for export to the south about four times as great as that available from an area worked by the traditional methods. Looked at from the landlord's point of view, the Border shepherds offered rents five to ten times as great as those they could hope to extract from their clansmen. From the point of view of the mass of the Highland population, the new methods meant that a shepherd and his dog and 600 sheep could profitably occupy an area that previously supported a crofting township of about 100 souls—in brief, it meant mass unemployment.

In England the population was rapidly multiplying in the industrial cities; at the beginning of the nineteenth century new industries and the Napoleonic wars were draining the countryside of men and forcing up the need and demand for meat and wool. The woollen industry, a principal contributor to national prosperity, was cut off from its Spanish raw-wool market and faced with disaster. Many Highland landlords shelved their sentiments and their clansmen together and accepted the sheepmen's rents and a pat on the back for patriotism. The Highlanders were cleared from the more fertile glens and hill grazings, cattle, ponies, goats and all. Some were moved by threats, some were moved by force, and some were left to face the cold and unrelenting wind of economics.

Many of the evicted crofters were re-established on poor ground on the sea-shore, bereft of extensive hill grazings. Crops and the sea gave them a living; tiny patches fertilized with lime-rich shell sand and seaweed and cultivated by hand with immense labour stood between these families and starvation. This was no place for a goat, and any goats retained on holdings of this kind would have to be tethered to preserve the crops. As we shall see in a later chapter, tethering undermines the goat's natural ability to avoid parasitic worm infestation, and almost invariably results in reduced productivity. Cows, more easily warded off crops and producing more milk per man-hour of herding under these circumstances, ousted the goat from favour.

During the more violent clearances, the greater part of the goat population must have been left to run wild on the hills. Domesticated habits always sit lightly on the goat, even the most inbred and highly selected of modern dairy strains being perfectly willing to turn feral* if given a suitable environment and opportunity. There must originally have been a great stock of 'wild' goats of this kind on the newly acquired sheep ranges. But goats commence their breeding season when the hours of daylight decrease and the hours of darkness increase at a certain critical speed. This critical rate of change is, of course, reached earlier in the year the nearer you go to the land of the Midnight Sun. Over the Highlands of Scotland the breeding season starts in July and August, and the kids are born in January and February to the feral goats—and only the very toughest and luckiest can survive the inhospitable welcome the Highlands offer at that season. Moreover, the goat does not like the wet, and in a climate such as that of the Scottish Highlands the feral goat population is very much limited to the number of dry beds available

---

* The term 'feral' is used here and in subsequent chapters to emphasize the fact that 'wild' goats in Britain are not 'wild' in the sense that red deer are 'wild'—they are, as the botanist would say, 'escapes'. The distinction is of some importance for the survival of 'wild' goats is not altogether dependent on their adaptation to their environment, but partly on a continuous supply of 'escapes'. Consequently the existence of these 'wild' goats is not a proof that goats can thrive under wholly natural conditions in this country.

on a wet night. Feral goat communities are still widely distributed in the Highlands of Scotland, but the numerical strength of each is fairly small and static.

Plate 1. Galloway Wild Goat with male kid.

Over wide areas of the Highlands and islands the crofters were not ordered or forced to make way for the sheepmen. Whether the slow attrition of poverty, hunger and discomfort was a kindlier fate than that suffered by the victims of the great clearances is doubtful; the result was very similar. Looking down on the ruins of abandoned croft houses standing among the head-high brackens that cover the old arable land, it is hard to tell today whether the desert was created by sudden violence or perennial poverty. The uncleared crofts collapsed through the interaction of two forces. The attraction of better-paid jobs in the south and overseas drew away the young and enterprising; the call grew clearer and more insistent as communications improved and the tales of successful emigrants multiplied. Those who remained behind were atracted by the relatively easy money that could be made, temporarily, by adopting the sheepmen's methods and marketing mutton and wool for a living. Subsistence agriculture is impervious to the ebb and flow of market prices; a pound of steak is a good meal whatever the price of store cattle. But once the peasant has gambled his fate on filling his pocket before his belly, he is at the mercy of the market. The market kills and blesses by turns, but cannot bless those it has once killed.

The goats on the uncleared crofts were directly affected by the call of emigration which drew away the human population and, in a different way, by the advent of sheep breeding and the change-over from subsistence agriculture to a crop-marketing economy. The tendency of the domesticated goat to turn feral, which is examined in detail in a later chapter, evinces itself at two seasons—at the onset of the breeding season in August–September; and at kidding time, if there is adequate natural feeding available. Kidding time with domesticated goats in the Highlands is best arranged to take place in March and April. Thus the tendency to turn feral, which requires careful and constant herding to prevent, occurs at seed time and harvest when there is a minimum of labour available to prevent it. The fall in human population released the goats to freedom and its perils.

The advent of sheep breeding in itself should scarcely have affected the goat population, for goats are good friends to the sheep and save many from death in bramble thickets and from falls from the cliffs. Yet there is a fable still prevalent among the sheep fraternity that goats on a sheep grazing will interbreed with the sheep and ruin the stock. The ram will serve the she-goat and the he-goat will serve the ewe; but only if frustrated in search of their own kind; very occasionally, as a rare phenomenon, the service bears fruit; there is a possibility that occasionally a ewe so served may go barren after an early abortion. On this gossamer of fact, which might support the delicate interest of a biologist, shepherds have hung an axe of execution. The goats on the Isle of Ulva (off the coast of Argyllshire) were slaughtered under this pretext some years ago. My own landlord deprecated my goatkeeping activities for the same reason. In the days when science and shepherds were strangers, many of the Highland goats went innocent to the slaughter.

The goat is by its nature the symbol and mascot of subsistence agriculture. It is first and foremost a household provider, and in this role its useful characteristics find their fullest expression. With the decline of subsistence agriculture the demand for goats declined. Under the pressure of a crop-marketing economy, the Highland crofters turned to cattle with some dairy qualities; with these they could go through the motions of eating their cake and having it. The calves were starved of milk to feed their owners, and then cashed in their stunted maturity. So the desperate hunt for cash deteriorated the cattle stocks, diminished the net income from the area and hastened the process of devastation.

When first the sheep invaded the Highland hills they throve on fresh pastures, almost free of internal parasites, and wintered on abandoned croft land still fertile from centuries of human labour. They sold to a roaring trade in a hungry market. The honeymoon was brief. By the later half of the nineteenth century, sheep farming paid so poorly that the Highland landlords were clearing off the sheep to give unfettered scope to the stalking tenant and his quarry. If sheep were a nuisance to the stalker, goats, the most keen-sighted of all our quadrupeds, were worse, their snorting whistle of alarm at the sight of a stranger carries farther than the sheep's. Feral goats still suffer from the stalker's rifle for this sin; domestic goats are still banned on some stalking estates. In the heyday of stalking, the cold war of today was too hot for a lot of goats. Here again the goats were preyed on by a fable: many years ago, on the shores of Loch Morar, a billy goat was run with the crofter's cows; this one had not the popular excuse of preventing contagious abortion; he was there to frighten away the red deer hinds. The idea is plausible: the stink of billy within a mile upwind would mask from the hind the more delicate aroma of an approaching stalker and undermine her sense of security; so she would move out of range – and a rank billy has some range. If the fable were popular, it must have killed some goats. I call it a fable on the authority of Dr Fraser Darling's observations of the deer and feral goats of An Teallach (*A Herd of Red Deer*, by F. Fraser Darling, Oxford University Press). It seems the hind's sense of smell is too selective, and its nature too inquisitive, for it to be so deterred.

With the exception of the influence of deer stalking, this tale of the decline of goatkeeping in the Highlands of Scotland covers the causes of its more gradual and piecemeal disappearance from the mountain districts of the North of England and of Wales. The change from subsistence agriculture to a cash economy; depopulation of hill districts; the tendency of the goat to turn feral if allowed free range without adequate control; and the goat's sensitivity to infestation with internal parasites when its grazing is restricted— these are the factors which have driven the goats from the hills of Britain. It is to be noted that while this exile from the hills is due in part to basic and irreversible changes in human society, it is due in part also to problems of management and husbandry which are by no means insoluble.

So far we have dealt with the goat population of Britain of one to two hundred years ago. The absence of accurate statistics of goat distribution at this period will not surprise the reader, who may be content to accept the evidence offered by local place names, songs and stories, and the statements of such authorities as Bewick, as to the approximate numerical strength of the goat. But now we have to admit that in the late 1970s, when the hens of Britain are numbered with annual accuracy, there is no authority who can state the number of domesticated goats in Britain to the nearest 5,000 with any statistical justification. The Ministry of Agriculture and the Department of Agriculture for Scotland hold statistics based on agricultural returns; that is, they can state the number of goats for which rations were drawn during the period of feeding-stuffs rationing. This represents anything between 10 and 80 per cent of the actual goat population, according to district. The British Goat Society and the Department of Agriculture for Scotland can also offer figures of the numbers of goats in each district which were served by subsidized stud males. While these figures cannot form the basis of any accurate estimate of total population, they are useful in indicating the ebb and flow of interest in goat breeding from time to time and place to place from the 1930s onwards. For the rest, our information must come from show reports, memoirs and articles written by the older generation of goat enthusiasts.

From the end of the nineteenth century until 1940 the main goatkeeping districts were in the North of England—Yorkshire, Durham, Lancashire, Northumberland, Cumberland—and parts of Wales, forming a natural geographical and economic area featuring extensive moorland and poverty-stricken industrial populations. The goats distributed their services almost equally between rural cottagers and such ill-paid industrial workers as the miners and railwaymen were then.

In the rest of England goats were more sparsely distributed, with local strongholds in areas of unspecialized smallholdings, such as Essex, and frequently associated with industrial villages.

The goat of those days was a shaggy creature of nondescript colour, yielding up to 4 pints (2·3 litres) a day. There were two main types—the Old English, with horns sweeping up, back and outward in a smooth curve, rather short on the leg, with a long lactation, giving milk with a butter-fat content of 4 to 5 per cent. The Irish was more popular in the hilly districts; and the type survives still in the west of Ireland and in many feral herds in England, Scotland and Wales. In the Irish type the typical horns rise straight and parallel from the brow, turning outward and a little back at the top in billies, remaining straight, pointed and business-like in the females. Leggier, with a shorter lactation, lower butter-fat percentage, and somewhat lower yield, the Irish goats were annually imported and distributed through the hill districts of Britain in nomadic droves from which the milkers were sold as they kidded. Until the First World War the Irish goatherd, shouting picturesque advertisement of his wares, squirting great jets of milk from his freshened nannies up the main street, was a regular harbinger of spring in the mountain villages.

Some of the feral herds along the west coast of Scotland are pure white. Local tradition attributes their origin to the ships of the

Fig. 3. Feral goat heads

The old billy on the left is of the type commonly found on the small islands off the Scottish coast and in the Border hills; the type is probably native or Scandinavian in origin, but local traditions describe them as survivors of the Spanish Armada. The head on the right is that of a typical Irish billy such as the Irish goatherds used to import annually into the hill districts.

defeated Spanish Armada which sought here a refuge they didn't receive. Certainly, most large galleons of Elizabethan days carried goats as a source of fresh milk and meat, but there is little Spanish about the type of these feral herds today. The Spanish dairy breeds are all coloured, and such of them as are horned carry the short twisted horn of *Capra prisca*. However, the Netherlands were part of the Spanish Empire in these days, and the white goat of Swiss Saanen type has a long history there. Perhaps these were the first Saanen importations. It is equally likely that they owe their origin to Scandinavian seafarers who frequented these shores, in whose homeland the white 'Telemark' goat has long been popular. The sea route to the Western Isles was assuredly more hospitable than the land route until the late eighteenth century. Many of these feral flocks exist on small islands. But goats are bad swimmers; the goats of Ulva were exterminated by being driven out on to a tidal reef when the tide was rising. The ability to swim a hundred yards would have saved them. So it is highly improbable that these island goats swam to shore from wrecked ships. The prevalence of the tradition that they did so suggests that their origin is wrapped in mystery and antiquity. If their existence were due to the obviously sensible practice of sending dry stock and males to uninhabited islands for the summer—to save herding them from the crops—then the mystery would not exist, and the colour of the goats would not be so prevalently white. It is tempting and not unreasonable to suggest that the Viking longboats, which carried cattle to Greenland in A.D. 1000 and pirated West Highland waters for centuries, may have carried the white goat of Norway to the Western Isles and their islet strongholds. It is altogether appropriate to believe that the Vikings sustained their heroics on a diet of goats' milk and kid. In any case, there can be little doubt that the native goat stock along the whole seaboard of Britain was liberally mixed with 'ship goats' from abroad.

The relatively poor performance of the native British goat stock was not, then, primarily due to inferior origins, but to lack of a coherent breeding policy among British goatkeepers. In most parts of Europe, native goat stocks have long been developed to a high level of excellence, especially where a measure of inbreeding and coherent breeding policy has been enforced by the isolation and essential discipline of tightly knit mountain communities, such as those of the Alps and Scandinavia. In Spain the guiding hand of the centuries-old Sheepmasters' Guild has aided the production of a number of distinctive, long-established and highly efficient breeds. There is evidence that exposure to low temperature is of assistance in developing the goat's capacity for food intake on which the efficiency and economy of goats'-milk production very largely depends. Lack of the stimulus has perhaps been a handicap to British and Irish breeders. Whatever the cause, the dawn of the twentieth century found Britain with a goat population not only small in numbers, but relatively low in quality.

The first steps to improve Britain's stock were taken by the founders of her Indian Empire. The ships carrying the cargoes of the East India Company bore homeward also executives of the Company who had become acquainted with the virtues of goats' milk and its giver during their sojourn in the East. For the sustenance of passengers *en route* it was customary for these ships to take with them a few goats from India, and to pick up replacement milkers at Suez and Malta. On arrival in England, these unfamiliar animals found their way to zoos or to the home of a retired Anglo-Indian, who had acquired a taste for goats' milk, often allied with an impaired digestion which would not function without it.

These new goatkeepers were men of enterprise and vision with that Victorian faith in human progress which invested hopefully and earnestly in all manner of speculations, animal, vegetable and mineral. Under the aegis of the retired Anglo-Indians and their friends, the goats of the Orient were crossed with each other and with the native stock. Swiss goats were added to the collection, and in 1879 the British Goat Society was founded. Even today, the management of the British Goat Society and of the leading pedigree herds remains substantially in the hands of the spiritual and real heirs of its founders. These were not men of the land: they might have had one foot in the paddock, but the other was assuredly in the city. Consequently the modern dairy goat has not been developed in the farmyard like the cow, sheep and pig, but in the paddock and the back garden where, by garden-scale methods, have been grown such of the foods of Britain's modern dairy goat as the corn merchant has not provided. The virtues of the 'improved' goat have not been selected for their value in the general context of British farming, but for their value in these highly specialized surroundings. Pedigree goats have no more been bred to take their place in the mainstream of British agriculture than have pedigree Siamese cats.

It is not intended to belittle the achievements of the pioneer and present breeders of the modern dairy goat, but to explain why their brilliant achievements have had so little commercial success or practical application. The magnitude of their achievement is hard to exaggerate. From a mongrel hotch-potch of goats with yields of 50 to 70 gallons (227 to 318 litres) a year they have built up six distinct breeds with annual yields averaging 150 gallons (682 litres) under fair management and 200 to 400 gallons (909 to 1,818 litres) in the leading herds. They have, moreover, succeeded in tying conformation to production so closely that few adult winners in the exhibition classes are absent from the prize lists in the milking competitions. Lactation periods have been extended to cover several years, and productive life up to ten years. All this has been achieved in about a century with the help of but a few dozen imported goats.

Before these achievements, and the productive potential they represent, could (or can) be developed into any great asset to national food production, two steps were necessary: first, to impress the improved type on the native stock; second, to discover the conditions and methods whereby the potential could be realized on any considerable scale within the context of British agriculture. We may win our battles on the playing-fields of Eton, but we cannot feed our millions on the produce of pony paddocks.

The U-boat campaign of the First World War gave the new goat its first opportunity. But numbers were still too small, and type insufficiently well established, for it to make any sizeable contribution to the national larder. Goatkeeping in general received a fillip, and more might have been done to improve the type of the national goat stock had agriculture been subject to any but the crudest organization. In the upshot the new goat found an extended patronage which survived the peace in a few more paddocks and back gardens.

The agricultural slump that succeeded the First World War slowed the development of the new goat, but didn't stop it. The general economic depression which followed produced the first notable step forward: on the one hand goatkeeping increased in importance in industrial villages and on the outskirts of towns in the depressed areas, to provide occupation for the unemployed and cheaper food for the hungry; on the other, official recognition of the new goat's value took the form of the Stud Goat Scheme, subsidized by the Ministry of Agriculture and operated by the British Goat

Society, whereby the services of improved males were made available to the goats of cottagers and smallholders at reduced fees. A parallel scheme operated in Scotland under the auspices of the Department of Agriculture. By the outbreak of the Second World War the ever-growing popularity of this scheme and the steady progress of the leading pedigree breeders had established a nucleus of about 5,000 pedigree goats, with a further 15,000 showing some degree of improvement.

Plate 2. R.46 Bitterne Tessa*. British Goat. Her remarkable fourth lactation was 12,393 lb (5626·42 kg) in 1,061 days, butter fat average 3·84 per cent. Photographed when running through for the second time. Twice winner of the Malpas Melba Trophy. Breeder, the late Miss Barnaby. Owner, Mrs J. Laing.

With this nucleus, the impetus of the national need and the whole agricultural organization of the country firmly under government control, the scene was set for the new goat to take the stage.

Male goat licensing would have brought about quick improvement in type; but the Ministry of Agriculture was already embarrassed with manpower shortages. Knowledge of goats and knowledge of agriculture were so seldom to be found together in the one head, that the War Agricultural Executive Committees did nothing of consequence to utilize the goat's potentialities to meet the national need. The ever-increasing acreage of cut-over woodland, which goats might have converted into low-cost milk, remained as a verminous fountain of weed seed to poison surrounding fields. With little more than 30,000 goats on the Ministry of Agriculture's records, the Ministry of Food reached the understandable conclusion that goat dairy produce was best treated as an insignificant anomaly. The British Goat Society, somewhat enfeebled by the absence of many of its most vigorous members on active service, fought a gallant uphill fight for the elementary privileges and food allocations that were automatically accorded to all productive livestock except goats. The goat's real potentialities for national service were never so much as pleaded.

The consequences for the long-term future of goatkeeping in Britain were, in some ways, catastrophic. Goatkeeping expanded, as it always must when dairy produce is in short supply. But it expanded in the artificial economy of the subsidy and price-control system, almost totally divorced from its true sphere of usefulness and economic justification. The national Goat Stock improved in type and productivity under conditions which alienated it still further from British farming.

The value of the goat lies primarily in its ability to convert to milk herbage that no other animal can utilize. Its suitability as the household dairy supplier enables it to cut the cost of milk distribution in some areas, and to utilize wastes available in quantities too small to be otherwise useful. Under the subsidy system, cows' milk was made available to all at prices below the cost of production and distribution; the welfare foods service made it available free or at a nominal price to a considerable section of the population, including children. Unsubsidized goats' milk became overnight a luxury food.

Goatkeeping declined wherever its previous justification had been economy. It was almost completely ousted from the industrial villages, retreating before the joint forces of subsidized cows' milk and rising wages. War work for women and labour shortage helped to drive the goat from cottages and smallholdings within the range of a dairy van. In general there was a retreat from the hill districts and the less well-to-do counties of the North of England and Wales, from all the strongholds where the first principle of goat management was maximum production for minimum cost in feeding; war economy spelt a new role for the goat as the supplier to a legal black market in dairy produce: unsubsidized, goats' milk, cream and cheese were also free of price control. The focal centre of the goat population shifted south to the home counties and the heart of the national black market.

No stigma attaches to the operations of goatkeepers under these circumstances. They did that which government regulations permitted and obliged them to do. A legal black market in luxury products has a definite social and economic value under conditions of war and economic stringency. Goat dairy produce and the other items that resided under the national counter solaced the jaded worker, met the unscheduled need and provided a valuable cornerstone to national morale. Though its potentialities were not fully exploited, the goat served a useful wartime purpose; it was not Austerity Britain that suffered by the arrangement so much as the goat.

The cost of milk production under these circumstances was not a matter of prime importance: ice-cream manufacturers were willing to pay 10s (50p) per gallon (4·5 litres) for goats' milk; fresh goats' cream sold wholesale at 10s per pint (0·6 litres). Animal feeding stuffs, though often scarce, were price-controlled and heavily subsidized—and this was a subsidy from which goatkeepers did benefit. Labour was the essential of goats'-milk production in shortest supply and steepest in price. The obvious and most tempting policy for the goatkeeper to adopt was to secure the maximum production for labour expended, irrespective of costs in general and the cost of feeding in particular. In practice this led to the maximum possible use of concentrates to obtain the highest possible yields per head.

Between 1939 and 1949 animal feeding stuffs were rationed. For much of the period the goat's concentrate ration was limited to about 1 lb (454 g) per day for a milker giving over six pints (3.3 litres), ½ lb (227 g) for lesser producers and a very small allowance for kids, to encourage early weaning. Towards the end of official rationing, admittedly, it was easy to obtain extra coupons; throughout the duration of rationing the local farmer would let the goatkeeper have the odd bag of oats on the quiet; but lack of concentrates, and especially lack of protein concentrate in winter, was a real headache for the goatkeeper for most of that ten years. Pre-war goatkeepers complained bitterly that they were unable to keep their goats in the style to which they were accustomed; newcomers to goatkeeping accepted the low level of concentrate feeding more readily, but everyone attempted to compensate by feeding additional unrationed bulk food of various kinds.

During these ten years of austerity, there was a good demand for goats; breeding was far from selective; flock numbers doubled; and the average productivity of pedigree goats increased by 11 per cent, with a 6 per cent increase in butter fat. In the next five years, the supply of concentrates was unlimited; the demand for goats decreased; breeding became highly selective; and flock numbers were halved. Yet average yield fell by 14 per cent, with a 12 per cent fall in butter fat. Had the breeders of that period retained their austerity feeding methods and used the worst available males, they could not have achieved such a catastrophic destruction of breed quality in so short a time.

Imagine an exemplary disaster like this befalling any other section of British livestock: the breeders would be in dismay, the causes investigated, the methods reformed. No such disquiet disturbed the goat fraternity (sorority?): they trotted on down the Gadarene slope for another five or six years before a lonely voice was heard to remark that 'things ain't what they used to be'. In 1965, with yields now 20 per cent below the 1948/49 peak, and still falling, some misgiving became generally apparent.

The technical reasons for the collapse of productivity are discussed in Chapter 7; but the reasons are not all technical. The fact is that from 1950 on, productivity had no great economic value for a majority of pedigree goat breeders. With cows' milk derationed and cows' cream legally on sale, goats'-milk sales ceased to be profitable; much goat's milk was diverted into the feeding bucket, for pigs, calves, chickens, pedigree cats and puppies, mink, silver fox—any consumer which provided some return without demanding an all-year-round supply, hygienic dairying or salesmanship. The profitability of such goatkeeping, if any, came to depend largely on sales of pedigree goats to other breeders, a good show record being worth more than a high yield.

The recent spectacular rise in the popularity of goatkeeping in Britain is revealed by the figures in Table 6 (p. 193). There are several reasons for this. In some areas cows'-milk deliveries have now been discontinued, forcing those with facilities to consider keeping goats. Families disenchanted with city life settle in country cottages with anything from half an acre (0·2 ha) to several acres. There is a general awareness of the higher level of nutrition and palatability derived from fresh, home-grown foods. Agricultural shows often include goat classes, in which beautifully-turned-out animals with well-filled udders catch the public eye. So the household goat, with her modest housing requirements and lavish milk yield in relation to her small stature, and her ability to make use of weedy corners and all the extras from the vegetable garden, has proved herself invaluable to those attempting some measure of self-sufficiency.

In addition to milk and its derivatives, male and any surplus female kids, milk-fed when milk yields are at peak, kill out well at three to five weeks for the table or deep freeze. The NPK rating (nitrogen, phosphorus and potassium content) of goat dung is exceptionally high—another asset, ensuring, through the compost heap, fertility for successive crops. Local goat clubs are doing good work finding reliable stock and starting these new goatkeepers, many of whom have no previous experience of dairy animals, on the right lines.

Table 6 also indicates the relative yields and butter-fat contents of the different breeds, as well as fluctuations in their popularity. A register recently opened for the Golden Guernsey is restricted to pure G.G. blood. These decorative, light-boned goats are listed among rare breeds, and it will be interesting to follow their progress in performance and popularity.

Stud fees have soared from £2 or £3 to £5 minimum up to £10 or £12. The increase is long overdue when one considers stud fees for dogs, £40 to £50 being commonplace. The Herd Book goatling bred for long lactation fetches from £80 upward, with the young milker fulfilling her potential making at least £100. Top show animals are valued considerably higher, and seldom come into the market at anything but a very big price.

Although the demand for milk far exceeds the supply, particularly in some localities, few commercial goat farms exist: perhaps three or four in England, and just one in Scotland. The market for hospitals and nursery schools should be fairly steady, but that for children suffering from eczema and asthma may vary widely in a particular place. Many grow out of these diseases at the age of seven

Plate 3. CH: Peggysmill Rachel Q*BR.CH. BT5774P British Toggenburg goat with consistently good butter fats. S. Peggysmill Bluster BT5584. D. Westcliffe Miranda BT5267P. Consistent winner in milking competitions. Classic wedge-shape conformation. Owner/breeder, Miss Rosemary Banks. (Photograph: BGS Year Book 1976)

or so, and are able to discontinue the goats'-milk diet, while other children in other localities then have to be supplied. Some health-food shops retail milk, yoghourt and sometimes cheese; but the perishable nature of the products and fluctuating demand make them scarcely a practical proposition for the smaller shops. Collection and distribution become increasingly costly, and there is not much profit for the farm. The 1978 price, ex-farm per pint, is probably only 14p to 15p, with the retail price from 19p upward to 25p or 26p.

Chapter Three

The Prospects for Goat Farming

When the goatkeeper looks at the potential market for his goats' milk and cheese, he will find three main outlets: customers whose needs are medicinal, health-food shops, and surprisingly, caravanners—the last usually in summer.

The customers of health-food shops, and the health-food counters of big stores, are usually the first fish to be hooked by advertisement of goats' milk. Though most of the regular goats'-milk customers will eventually be those with real medical need, the health-food enthusiasts are a providential receptacle for the summer milk surplus of the goat-dairying specialist. Regular customers must have an all-year-round supply; goats'-milk production being inevitably much lower in winter, there is always a surplus in summer. To a small extent this may be used to build up a store of deep-frozen milk for winter use, but most of it can be sold to the health-food shops in the form of genuine, live, goats'-milk yoghourt, which is prepared by an exact, but not at all laborious or expensive process. You cannot sell this really delicious article through the multiple dairies, because it would make a mockery of the often insipid product which they make themselves and sell as 'yoghourt'.

In less densely populated areas, it becomes increasingly difficult to find enough goats'-milk customers within reach of a daily milk round to provide the dairy-goat specialist with a reasonable living. Co-operation with medical authorities, and delivery through the retail dairy network, are harder to arrange. Publicity and advertising demand the expenditure of more effort and receive a smaller response. Somewhere along the road from the city to the mountains, the need for fresh goats' milk must be met by part-timers—either by domestic goatkeepers selling a constant surplus, or by a goat-dairying sideline on a mixed farm.

The producer-retailer of cows' milk is ideally situated to run a goat-dairying sideline; he already has the dairy equipment and the means of delivery and publicity available. A regular proportion of his cows'-milk customers will be suffering from indigestion, insomnia, unhappy bottle-fed babies, or other such distress that goats' milk can allay. They will accept *his* advice to buy goats' milk; they will be glad to pay a little extra for the relief it brings.

Many producer-retailers do a line in cream. Goats' cream offers pleasant surprises to producer and consumer. Being richer in butter fat, on the average, a gallon (4·5 litres) of goats' milk yields more cream; goats' cream being of lower specific gravity, a given weight takes up more space; a gallon of goats' milk yields cream to fill 20 per cent more cream cartons than a gallon of cows' milk. The housewife will find that the goats' cream whips to greater bulk, 'goes further', and does not give grandpa indigestion.

The problem of the summer surplus of goats' milk on a mixed farm disappears profitably down the throats of caravanners, piglets, calves and fattening chickens, releasing more cows' milk for sale. Calves, especially Jerseys and Guernseys, do better on goats' milk than on the milk of their mothers; they tend to scour less frequently. One prominent goat breeder used to rear the most promising heifer calves for a leading breeder and exhibitor of pedigree Red Polls.

The cow and goat dairy farmer must resist the temptation to mix the milks. A trace of cows' milk in a goats'-milk diet can wreak havoc on an allergic patient. The addition of goats' milk to a bottle of cows' milk diminishes the depth of the cream line on which milk quality is commonly judged. The fat of goats' milk is so finely divided that it takes several days to rise and form a cream line.

Much of the goats' milk available in the countryside must come from the regular surplus of domestic goatkeepers, and the show flocks of goat fanciers. The economic pros and cons are not the decisive factor in domestic goatkeeping. It is more a matter of social values, of the accepted way of life. Goatkeeping is often a symbol of a deeply felt minority faith. Between those who work overtime in an uncongenial occupation to buy labour-saving domestic gadgetry and manufactured pleasures, and those who cut the cost of living by the (to them) congenial labour of goatkeeping, gardening, and doing-it-yourself, there is fixed a great gulf. From whichever side you view it, Hell is on the other. Those on the goatkeepers' side are a minority, but an increasing one.

In the rural areas it is often impossible for the doctor to prescribe goats' milk without prescribing domestic goatkeeping to produce it. As household providers, goats can perform their function under an extremely wide range of conditions—the world milk record was broken by a goat fed largely on greengrocer's waste in a Bermondsey back street in the middle of the V-bomb blitz. The retail dairy's costs and profits are increasing very much more rapidly than those of the milk producer, so that the economic advantages of home production are steadily improving. Widespread pasteurization has depressed the palatability of available cows' milk, so home-produced goats' milk not only saves money, but tastes better too. If that is what the doctor orders, who can say no?

At times it may happen that the cost of goats'-milk production is not a very serious consideration; when dairy produce is scarce or money plentiful, the special qualities of goats' milk can always secure a market for a regular and clean supply, with little regard to price. But the market for agricultural produce sees bad times as well as good. If goat dairying is to enjoy any useful degree of stability, and any resistance to adverse economic conditions, it must be founded on a supply of cheap fodder.

Probably the ideal site for goat farming is a block of cut-over woodland within marketing range of a fair-sized town. The second growth should be at least two years old, and preferably in that stage when thistles, nettles and docks are hurtling weed seeds into the surrounding fields from behind a barricade of brambles and saplings. Stocked at the rate of one goat to the acre (0·4 ha) such ground will sustain yields of 6 to 10 pints (3·4 to 5·7 litres) a head from leaf-bud to leaf-fall, and winter yields of 2 to 3 pints (1·1 to 1·7 litres) a head with little or no supplementary feeding of bulk foods. During the winter, the goats will be living mainly on tree bark and brushwood; if the stocking is any heavier than one goat to the acre, the cumulative deterioration of the woody growth will ruin the goat grazing in about ten years. But if the goats are hand-fed with roughages during the winter, the summer stocking capacity can easily be raised to two or three goats per acre, depending on the size of the goats.

The nature of the sapling growth is also important. Elm, ash, hazel, willow, hawthorn and sycamore are the best for summer grazing; elm, willow, ash, holly and ivy are the most valuable in winter; oak is of limited value; so are alder and birch; beech is virtually useless. Of the shrubs, briars are most valuable in summer, brambles in early spring and winter, gorse and broom in late winter and spring; heather is first-rate feeding from autumn to midwinter, and blaeberry (whortleberry) during spring and summer.

Mature woodland also makes good goat grazing, if of suitable type. Softwoods are valueless, but a stocking of one goat to three or four acres is suitable for mixed hardwoods; though goats will put paid to natural regeneration, they will not do substantial damage to mature trees. Once the leaves have fallen and started to rot, mature woodland will not feed goats through the winter unless it has some of the winter-feeding shrubs mentioned above.

In view of the afforestation programme, it is perhaps narrow-minded to look at cut-over woodland simply as a source of goat milk. Foresters would rather regard it as future planting ground—but as such they cannot but view it over the top of a more or less insurmountable obstacle. The capital and labour required to kill the second growth is great, and the supply of both is scarce. Goats will kill it by barking the saplings in the winter, and fire will clear what they kill. Blocks of 100 acres (40·5 ha) will support an economic goat dairy unit for a man and a half; over larger blocks, systematic destruction would be more difficult to organize, and the control of the goats might cause trouble. Fencing off an extensive area of cut-over woodland into 100-acre blocks might prove expensive, and the goats would take perhaps ten years to achieve their purpose; but schemes more expensive of land and money are known to forestry.

As a sore in the eye of the travelling farmer, few sights can compare with that of the hostile arsenal of weed and vermin in a cut-over wood. But occasionally he passes a comparable menace in worked-out clay pits and opencast iron mines. Here, too, the goat can control, suppress and convert to milk the troubles of the general farmer. Moreover, the amount of fencing required on such sites is relatively small; though able to negotiate anything which is not quite smooth and vertical, the goat is retained by a sheer rise of only 5 feet (1·5 m) and by a sheer drop of 10 to 15 feet (3 to 4·5 m).

In coastal and hill districts, many a farm has a considerable waste acreage occupied with cliffs and rocks and a vegetation consisting mainly of brambles and briars and a certain amount of dangerously situated grass. This is useful goat grazing, however inhospitable the rock may appear; but it will not support so many goats as heavy scrub.

Under domestic or commercial farming conditions, goats will do exceedingly well fed on horticultural wastes. Not only will they consume and cash the very wide variety of garden wastes, from fruit-tree and shrub prunings to dock and nettles, but will provide a good supply of manure of a kind highly suitable for gardening purposes, being light to handle, warm and quick-acting and almost entirely free of weed seeds even without being composted; goat dung made during the summer has a higher nitrogen content than horse dung, cow dung or sheep dung; weight for weight, it has more than three times the value of FYM. As an adjunct to horticulture, yarded goats can often be fitted into the existing organization of the holding with a positive saving of labour and materials. Whether the milk is marketed at a good price for human consumption or used for stock rearing on the holding is not then of prime consequence.

The stock returns for horticultural holdings frequently include a few pigs. For piglet feeding, the high digestibility of the fat and protein of goats' milk and its superiority in mineral and vitamin content give it the same kind of advantage over cows' milk that it has in infant feeding. Pigs and goats always make a happy combination which avoids marketing difficulties for goat dairy produce if required, but can also accompany a sideline in goats' cream, which can be sold to compete in price with cows' cream.

It is an attractive thought to employ the goat, which can utilize the rougher fodder with such unique economy, to make milk of moorland and hill pastures. But the notion is only occasionally practicable. Apart from the difficulties of marketing goat dairy produce from remote areas, the herbage of hill grazings and moorland is often unsuitable for goats. Where the soil coverage of the hill is predominantly free-draining peat, and the vegetation consists mainly of heather and other ericaceous plants such as blaeberry (whortleberry), it will sustain moderate yields of 4 to 6 pints (2·3 to 3·4 litres) per head, and good health. Heather is good winter feed, is rich in lime and has a respectable content of protein; but it must be burnt regularly to keep it productive. Blaeberry is first-rate summer feeding for the goat, being mineral-rich with a good protein content, but it does not flourish among heather which is regularly burnt. Hill pastures including substantial acreages of blaeberry as well as well-burnt heather are relatively rare, and are the only ones really suitable for commercial goat dairying—and then only if they are within reach of a market.

The hill grasses such as white bent and molinia are minerally deficient; the goat has small appetite for them, and is liable to mineral deficiency diseases on such a diet; moreover, these grasses grow mainly on sodden ground, which goats instinctively dislike, and which harbours a wealth of parasitic worm larvae peculiarly dangerous to the goat.

Nevertheless, as an adjunct to a hill sheep farm, a few goats will usually pay their way. The hard-earned hay from marginal meadows pays poor dividends if fed to a house cow which is dry for two or three months of the year. The same hay will sustain two or three goats with an all-year-round yield of greater volume. On a hill sheep farm a few goats can usually find a sufficient variety of shrubs, trees and weeds to sustain their summer yields, without destroying what shelter facilities the limited woodland provides.

Goats' milk is very much better digested by the lamb than cows' milk; it can be fed undiluted and fresh from the earliest stages. Fed to Blackface ram lambs, goats' milk ensures the production of classic horn growth; cows' milk produces too close and light a growth, while the milk of a well-fed ewe of good milking capacity produces a growth which is too wide and heavy for fashionable taste.

Goats will often prove useful in keeping sheep out of dangerous cliffs in the early spring, by clearing the rock ledges of green temptation before any other beast can get near them. On very high and extensive cliffs, the cure will not be complete, as the sheep are inclined to follow the goat tracks down the cliff, and cannot follow them up again. But provided the cliffs are sufficiently restricted in extent and vegetation, to enable a reasonable number of goats to control the growth on them in early spring, the cure is completely effective.

Goats are also extremely useful to the sheep farmer seriously troubled with brambles. If tethered, preferably on a running tether during early summer, goats will graze the brambles to extinction. The job is not suitable for a well-bred dairy goat, who will become half-starved in the process; but a rough scrub goat will do the job effectively with little discomfort to herself. Goats of a milkier type can still play a substantial role in bramble clearance, as they will happily graze to destruction any brambles on their range which have been scythed in the previous winter; they will not tackle an untended bramble thicket on their own, though they may stop it spreading.

Any hill farmer who has expended time and energy on bracken clearance will be familiar with the distressing phenomenon that a heavy growth of brackens, once controlled, is usually succeeded by a fine crop of nettles and thistles and docks, which not only limit the grazing value but tear the wool of the sheep. This is a nuisance which will not survive a year of goat grazing, but tends to get worse under sheep and cattle grazing.

The hill farmer does not usually look for a dairy round as a source of income; but the house cow has often helped the economy of the holding by producing a surplus which is taken up by neighbours whose cow is dry, and by summer visitors. With the price emphasis and subsidy on beef cattle, this surplus is rapidly becoming a distant memory in hill districts. A small goat dairy herd will find ready business in the neighbourhood of many a remote holiday village, and will pay its way on the hill farm which specializes in summer visitors. People will form an illogically rosy memory of the beauties of the holiday resort where they were relieved of the minor digestive distresses that goats' milk so efficiently allays.

Goats have occupied a minute corner in general agricultural practice for many years. It is worth recording that they are still entitled to it. Science has not yet put paid to the unlikely belief that a goat run with a herd of cows will prevent or reduce the incidence of abortion. Sympathy is certainly due to the veterinary surgeon who complains that it is bad enough dealing with a byre full of aborting cows, without having the smell of billy goat to add insult to injury. Goats have no effect on contagious abortion of cattle. But the presence of a few goats with the herd is the best protection available to many farmers against the danger of the cows aborting as a result of ergot poisoning.

Ergot is a fungus which grows on the flower and seed heads of cereals and some grasses, especially ryegrass. Perhaps because the flower and the seed heads are richer in minerals than the rest of the plant, perhaps because they are less liable to harbour parasitic worm larvae, a goat grazing grass shows a marked preference for this portion of the plant. In a field of flowering ryegrass goats will each eat nothing but the flower heads, ergot-stricken or not. Goats do suffer from ergot poisoning, and the goat's grazing habits give substantial protection to the cows at slight risk to the goat. Billy goats are preferred for the purpose, for obvious reasons.

Goats have a comparable corner in the stable; but here sex is immaterial and the purpose is different. There is an immense amount of empirical evidence, but no scientific explanation, for the belief that the presence of goats in a stable prevents the disease known as 'staggers'. Such a notion is as scientific as any until it has been scientifically disproved; whether it is acceptable to the scientific horseman or not, the goat earns its place in his stable as an insurance against fire damage.

Horses naturally rely on the speed of flight to escape from enemies and perils; the goat's safety lies in strategic retreat to inaccessibility. Panic, which lends wings to the horse's flight, is foreign to the calculated retreat of the goat. On several occasions during the war the truth of the old tradition was demonstrated afresh. The goat's calmness in the presence of fire and high explosive enabled it to lead horses, otherwise unmanageably panic-stricken, to safety.

Goats are used extensively in Austria, Italy and Spain to consume the organic wastes of several industries—including the important olive-oil and tomato-canning concerns. These countries are not so burdened as Britain with legislation affecting the disposal of wastes, and the subsidiary goat enterprises are run at a profit. In this country it might well prove more economic to feed wastes to goats at a considerable loss, rather than to dry and burn them in compliance with regulations. Edible wastes of low fibre content already have a capacious dumping ground in the national pig trough. It is the high-fibre wastes of the canning industry and the fruit-syrup factories which appear to provide the most suitable goat food. To the enterprising management advice from the agricultural advisory service and the British Goat Service is free; 'goat girls' are available, and they cost less than engineers. But in the trial stage it would be unwise to trust too implicitly to Continental experience—the type of goat used on the Continent for this purpose is, for the most part, the small goat of *Capra prisca* ancestry which seems to be able to thrive on a more concentrated ration than our *Capra aegagrus* derivatives. In these countries labour costs have in the past been of less account, and a system of individual stall feeding is adopted; here a yarding system, as described in Chapter 5 would appear more economic in labour, and is certainly healthier for the goat.

Chapter Four

The Control of Goats

Much popular prejudice against goats, and many real problems of goat farming, arise out of the special difficulty involved in keeping goats under control. In cultivated areas, the goat's contempt for the normal stock fences, and its destructiveness to trees, gardens and crops are notorious; on waste land—mountain, heath and cliff—the tendency of the goat to return to the wild (turn feral) is a significant nuisance. Even when narrowly confined to a yard or loose box, the goat contrives to trespass on the privacy and liberty of her owner by refusing to thrive or be silent out of her owner's company. For all these sins the penalty and solution may be found in the collar and tethering chain—the most laborious method of controlling farm stock, and one to which the goat is naturally ill adjusted.

It must be admitted that the goat is inherently less easy to control than other members of the farmyard community; yet the nature of the goat is disciplined, co-operative and intelligent; most of the difficulty in controlling her arises out of the fact that the goat's psychology, her requirements in the way of food and shelter, and the specifications for the fence that will control her, are so very different from those of other farm stock.

Psychologically, the goat has some particularly striking peculiarities. It is almost impossible to drive goats; they do not share with the other grazing beasts the convenient habit of packing and turning their tail to the approach of danger, dogs and omnipotent man; nor do they share with cattle and horses the instinct to put as much distance as time will permit between themselves and an object of suspicion. Goats scatter and face the enemy that comes suddenly upon them, committing their safety, not to speed and distance, but to superior agility and manoeuvre. The pursuit of a frightened cow or horse follows a bee-line, which can be directed by outflanking; the pursuit of a goat follows the course that the ragged rascal ran—round and round the rugged rock.

It is certainly possible to perform the illusion of driving a flock of goats, but only when there is a personal and private arrangement between the goatherd and the flock. In reality the flock is being led from behind, as the king billy of the wild flock may often lead; if the goatherd who is 'driving' the flock turns about and goes home, so does the flock.

The unfortunate Mary was able to call the cattle home across the sands of Dee, so long as her call meant relief to the cow with a distender udder or food to the cow with an empty belly. To cows with full bellies and empty udders Mary calls in vain, and has to send the dog to fetch them. But a goatherd can call his flock of hungry goats away from their foraging, and they will follow him, complaining but obedient, to be shut up in a cold yard for the night.

Yet should Mary let her cattle roam the sands of Dee with their suckled calves from year's end to year's end, and never set eye on them, she and the dog can still drive them home when required with little more trouble than usual. Let goats roam unattended and unrestricted with their suckling kids for a few weeks, and, if you see them again, they will be wild creatures; there is a slim chance that if you follow the flock slowly and call to them in familiar tones they may eventually respond; but as for using a dog on them, you are as well advised to send him after the red deer.

The psychological relationship between man and his domesticated animals, horses, cows, sheep, dogs, and poultry, is specialized. The behaviour of these animals towards man is different from their behaviour towards any other creature, and different from the behaviour towards man of their wild counterparts in captivity. The relationship is the product of centuries of selective breeding, and may be said to constitute their domestication. In this sense goats are not domesticated. At times they treat man as they treat members of their own species, as other times they treat man as their wild ancestors do in a natural state. There is no qualitative difference between the behaviour towards man of wild goats captured young and that of so-called domesticated goats.

Anyone who has the care of goats soon grows to realize that the relationship between the goatherd and his flock is a great deal more personal, more intimate and more delicate than is usual in the farmyard. It is in fact similar to the relationship with a gregarious animal that has been tamed from the wild. But those who tame and control wild animals usually have an advantage that goatherds often lack—a good knowledge of the social behaviour of the animal they are taming. It is a great deal easier to control goats if you understand something of the social structure and routine of the wild flock on which the behaviour of domesticated goats is based.

The wild goat flock consists of up to thirty or forty goats of all ages and both sexes. In structure it is an easy-going patriarchy. The king billy of the flock rules by right of strength and courage; he maintains discipline, keeps the peace and coherence of the flock and is its constant guardian. He leads the single file in peaceful movement, but shares the practical leadership of the foraging expedition with an old she-goat, the flock queen, who will normally outlast a succession of kings. Billies are expendable; their great size, ever-growing horn and hair and extravagant sexual habits throw a constant strain on them; their fearless readiness in defence of the flock secures for them an early and heroic end if the opportunity presents itself. Yet the billy is utterly egocentric: the flock is for him an extension of his own person; his care for the flock is but arrogance, and the flock but render him his due. They accord obedience while he is present; but they do not lament his absence. It is the old she-goat, the flock queen, who is the mainspring of the life of the flock.

When she stops to feed, the flock feeds, when she raises her head from browsing and stares at the billy, the king moves on to the next foraging site. In the face of danger she leads the flock to safety, while the billy brings up the rear and holds the enemy at bay. She sounds the alarm if any member of the flock is missing, and will not be content until he is found. When she is absent the flock is in turmoil. If the flock grows too large for its range, it is an old she-goat that leads the breakaway party, with a young billy as guardian, to form a new flock.

The flock is coherent throughout the greater part of the year. Most of the kids are born within the space of a few days, and at this time the goats in kid break away from the flock for about a week, the king billy keeping the remnant together till the flock re-forms. For some weeks thereafter the flock queen is too much occupied with maternal duties to take much part in flock leadership, which rests almost entirely on the shoulders of the billy; at this season, when his protection is most needed, he enjoys more power than at any other.

About a month before the onset of the breeding season, the king billy is much occupied in chastizing his sons and grandsons, the mothers cease to call their kids and the flock travels long distances on to poorer grazings in a very straggly formation. Only at nightfall does the flock re-form during this period; it re-forms where its wanderings find it, and does not return to its citadel for the night as it does at other seasons. This late-summer wandering serves to wean the kids and alter the flock's diet from the soft, milk-producing forage of summer to the hard, heat-producing forage of winter.

When man enters the social circle of the goat flock, he (or she) assumes the rank of kid, flock queen or king billy, according to circumstance. As kid, man can milk the flock peaceably; as king billy or flock queen, man can lead the flock. If the goatherd has nothing to do but milk and lead his flock, he will have no more difficulty in controlling his goats than the little boys, of seven to twelve years of age, who keep their flocks from trespassing in the surrounding fenceless fields all the way from Lisbon to Pekin. There is no great difficulty in getting into the social circle of the goat's confidence; the difficulty arises in getting in and out at your convenience and still exercising control.

Many goatkeepers start with the indiscretion of weaning their kids on to a feeding-bottle at birth, and keeping them separate from their emotional mothers until the fount of maternal milk and affection has dried up. The mothers so treated expend part of their energies, properly devoted to making milk, in calling for their human kid-substitute; and devise whatever means and mischief their ingenuity can contrive to bring their kid-substitute hurtling out of the house to chase and handle them. It is more peaceful and profitable to allow kids that are to be kept to suckle their mother for the first four or five days, then wean the kids on the bottle, but leave them in small kid-boxes, besides their mother in the goat-shed, where they can be seen and smelt and heard, but not suckled, for a further ten days. Thereafter they can run with their mother without fear of their sucking her. Kids make good progress if allowed to suck their mothers on range. Contrary to popular opinion, kids so reared are more manageable when adult than kids reared away from the mother; emotional deprivation in youth does not make for emotional stability in adult life in man or beast; the goat has a deeper fund of patience, time and topical knowledge with which to educate and discipline her own kid, than any human mother-substitute.

The policy of allowing the natural social organization of the goat flock scope to develop is, in general, a sound one. The flock of intensively kept goats devotes itself more wholly and happily to milk production if it is allowed a modicum of communal life; yarding goats is preferable to stall-feeding for this reason among others (always assuming that the yarded goats are all dehorned or hornless). Unless the goats have opportunity to give a practical demonstration of their strength, agility and character to one another, they lack the data on which to appoint a flock queen. If there is no flock queen in the flock, her human attendant becomes for each member of the flock an indispensable companion in whose absence she feels insecure and ill-content. As goats are highly intelligent, they soon learn that the more noisy, destructive, wasteful and faddy with their food they are, the longer they enjoy the desired company.

To keep one goat alone, unless you are prepared to spend most of your life with her, is not only troublesome but extremely cruel.

If the goat is permitted to love its own kid, and the flock to embody their sense of security in one of their own number, it becomes possible for the goatherd to enter and withdraw from the flock without disturbing the even tenor of its way. The goatherd then fills the role of kid only briefly at milking times, and for the most part occupies the position of king billy in the social structure of the flock.

Even so, withdrawal from the flock is not always easy. It is easy for the goatherd to leave the goat shed or field and shut the door or gate; provided he leaves kids and a flock queen behind him, he does not leave a bleating vacuum. It is less easy for the goatherd to lead the flock out on range and to leave them there.

The king billy leads his flock in peaceful movement, with the flock queen close behind or beside him. Both know the rounds of their range in detail; according to wind and weather and season they follow a predestined course. King billy stops at an ash thicket; the flock queen starts stripping the young leaves; the rest of the flock follow suit; though other palatable forage be close at hand, ash leaves are the *table d'hôte* for all. King billy reaches highest and eats ravenously; flock queen bends down the branches for the smaller fry, but doesn't forget her own needs. After about ten minutes the flock queen stops eating, approaches king billy, who is still stuffing himself, stares at him pointedly and, if necessary, makes a small remark. King billy stops eating and moves on. The rest of the flock who have scattered through the thicket, hurry to re-form the single file and follow. Next stop is for a patch of scrub oak (to balance laxative ash), where the same procedure is repeated. After perhaps a dozen stops and starts, the flock settles for the day on a suitable area of mixed browsing and there each one eats to repletion, lies down to cud and eats again. Towards evening king billy leads the flock back to their 'citadel' or goat shed by stages. Occasionally the flock queen or another senior member of the flock may stop for a moment to satisfy some personal requirement—say conifer bark—king billy waiting politely or holding on his way according to the importance of the goat who has stopped; but the rest of the flock does not feed.

The goatherd who knows his flock and range can lead on the first stage or two of their wanderings, and as soon as they are all busy at a popular stop he can quietly slip away. The flock queen will call him when she is ready to move on, but if he has disappeared she will take the lead herself. This tactic works well in broken wooded country; but in open country the delinquent leader is caught in the act of escape and the whole flock comes scampering after him.

Under such conditions the goatherd must use plain language to tell his flock when he has ceased to be their king billy. Mere rudeness won't suffice; king billy's manners are not of the best; a push and a grunt and a show of displeasure convey his warning to the flock to obey and follow him more closely. But king billy doesn't throw sticks or stones; the performance of such an act is a special characteristic of man and monkey which is peculiarly repulsive to all other species. So long as it is carried out calmly and ceremonially, so as not to be mistaken for mere rudeness, the flock will turn their backs on the goatherd, will go their way and leave him to go his. The popular use of effeminate males has produced flocks of half-witted goats which may be incapable of finding amongst themselves a flock queen of sufficient independence of character to accept even so broad a thing as the thrown stick. If such a flock refuses to be content without a king billy to lead them, the only cure is to give them a real one.

But the introduction of a male goat into the flock on range brings into close perspective a problem that is inherent in the relationship between the flock and the goatherd. Because the only way of controlling the flock is to be accepted as a member of it, the goat-herd's position is always open to the same challenge as that which faces the flock queen and the king billy. Occasionally the flock queen may challenge his right to lead; an adult male with the flock on free range will always make the attempt at least once. Honeyed words and bribery are wasted on a rebellious goat. A good sound trouncing is needed; not just an angry slap, but a stand-up fight and a walloping thorough defeat.

As a prefect at school I was told to inquire into the circumstances whereby a certain small boy, of peculiarly retiring character, came to get his Sunday-best suit covered with long white hairs and peculiar odour each Sabbath day. I followed the small boy when he took the liberty of a Sunday-afternoon walk by himself. He led me several miles across country to a small paddock in which there was tethered a very large white billy goat with awesome horns and ferocious habits. The billy must have weighed twice as much as the small boy, who approached him purposefully. Having taunted the billy to the end of his tether the boy grabbed him simultaneously by one ear and the tail, drew the two extremities together, forcing the billy to stagger round in a circle, downed him, rubbed his nose in the dirt, and then let him go. Then he did it all again. Satisfied, he returned to school and his retiring habits.

For the treatment of repression in small boys, better techniques may have been developed—but in seventeen years of goat husbandry I have been unable to discover a technique for the treatment of rebellious goats which is as effective, nor one which calls for less strength in the operator, nor one which is less painful for the goat. The need for some such treatment is widely acknowledged; for lack of a humane and practical one, recourse is had to hunting crops and other inhumane and impractical measures.

Not that the male goat is normally obstreperous—on the contrary, a vicious billy is a rarity, and his ill nature due either to being tethered and teased by small boys and dogs, or to being closely confined and handled with fear and distaste by attendants unable to control him.

In the wild flock the king billy is one of the most tolerant patriarchs of the animal world: he puts up with immature and adult males; personally instructs the youngsters in good manners and flock defence; and only hunts them off during the height of the breeding season—and then with more formality than effect. Young males happily accept the lead of the king billy, and are equally prepared to accept the goatherd as his substitute. But an adult male, though normally amenable will, if run with a flock on free range, challenge the goatherd's lead at two seasons of the year.

During the period which follows the birth of the kids, when the flock queen is occupied with maternal responsibilities and the billy's defensive potentialities enhance his importance in the eyes of the flock, an assertive male will often try to take control out of the goatherd's hands and to turn the flock feral. But the call of the milking pail has so urgent an appeal to the freshened goats, that, though the goatherd may have to go and fetch the flock for milking, there is little fear of losing control when he finds them.

But in the month before the onset of the breeding season, there occurs an unfortunate community of purpose between the self-assertive male and the milker who is trying to wean the milking pail. In wild country there is a very serious danger of the flock turning feral, whether the billy runs with the flock or not; the danger is greater if he is with the flock.

The control of goats is mainly a matter of preventing them from trespassing, and in wilder districts from turning feral. The control the goatherd can establish is effective so long as the goatherd is present, and, in so far as he can inculcate a routine, it is effective when he is not present. For a free-range herd, the ideal is to establish a series of circular tours, each appropriate to a particular season and weather, and none of them making direct contact with forbidden ground. The initial training takes much herding time, but once the circular tours are established, the flock will not place a hoof outside its territory unless the goatherd leads.

For nine years we herded up to thirty-two goats on free range over 1,000 acres (405 ha); some 300 yards (274 m) from the goat sheds a colourful garden lay unprotected and easy of access under their eyes each day. Only once was it raided—by a 'new boy' in the flock; it just happened to be off their beat. So did a patch of brambles which were a death-trap to sheep, a month of regular herding was needed before the goats would include it in their self-conducted tours. On the other hand, a wartime goatkeeper in Kent kept his goats in a large paddock separated from his garden by a thoroughly goat-proof fence. During a daylight air raid a plane dropped a stick of bombs across the holding: one in the garden, another on the fence between garden and paddock; while the whine of the departing plane was still in his ears, the owner poked his head out of his shelter to see what had become of the goats grazing in the paddock. They were eating cabbages in the garden. The garden was on their beat—despite the fence.

The minor problem of preventing goats from poisoning themselves only arises when the basic principles of goat control are ignored. Discrimination in matters of diet is a dim instinct in goats; education on the subject is a main feature of their social life. If the social life is stunted, by lack of communal activity and separation of mother and kid, not only are the members of the flock particularly liable to poisoning, but they have a jittery approach to all food, wholesome or not.

Safety from poisoning in the natural flock is achieved by the discipline of the foraging expedition, which demands that every goat eats the same kind of herbage, in the same patch and at the same time as the leader of the flock. Though most of the older goats may liberalize this discipline to some extent, it provides a constant fount of instruction as to what may be eaten, and where and when. In a flock with the active communal life, it is amusing to test our reactions to unfamiliar food, by leaving some unfamiliar but acceptable article, such as red cabbage, in their way. You will be rewarded by a grand display of ham acting.

The first goat to spot an unfamiliar food stops as if shot, snorts, and adopts the tense stance with feet set wide and ears pricked, exactly like a red deer hind who has caught a whiff of stalker. Then comes a cautious, inquisitive approach, nose stretched forward and the body bunched for lightning retreat. Then the flock queen takes over the investigation, adopting the same poses, and the rest of the flock huddle and watch. First the flock queen noses the object; then she nibbles a bit and spits it out; nibbles again, and chews tentatively; if good, she swallows it, and, as if revelation had suddenly come upon her, proceeds to wolf it. The other member of the flock, after a cautious sniff of identification, follow her example; each secures a morsel at least. If the mysterious food is not good, the flock queen spits it out with vehemence, snorts, fusses, runs frantically about wiping her mouth on the grass and pawing at it, and generally puts on a superb act of frightened disgust; then, having done justice to her station, she reverts to her normal foraging with sudden equanimity. Each member of the flock goes through the pantomime—the billy snickering his protective concern. When the

others have finished their pantomime, the billy engages the offending herb in heroic combat, and may take half an hour in obliterating
its presence: a memorable performance which must imbue the flock
with a great sense of digestive security.

The only aid the goatherd can lend to the natural discrimination
of the flock is himself to draw attention to, spurn and destroy any
poisonous weed which evades the notice of the foraging flock.

Goats generally have good road sense: a little on the panicky side.
Here, their reluctance to run directly away from an approaching
menace is serviceable; they leap to the bank or the road fence and
never run on in front of a car as sheep and cattle will. But a youngster
may miscalculate the speed of approaching vehicles and try to cross
the road to join companions on the other side. So all the goats
should be kept to the one side of the road; if tethered on the
roadside, the length of the tether should not permit them to reach
the roadway. As goats grazing roadsides are not bona-fide road
users (travelling from one place to another) they have no legal right
there, and there is no redress for accidents, in law.

If these comments on the psychological aspects of goat control
have been mainly negative—in the sense of preventing the goats
from doing damage to themselves or others—it is because the
positive job of maintaining a contented and productive flock is very
much the same for goats as for other stock. Regular, peaceful
routine is, perhaps, even more necessary to the naturally disciplined
goat than to their more anarchic farmyard companions.

The normal routine, which the goatherd aims to inculcate, is for
the goats to forage purposefully and peaceably from their feeding
racks, fields or free range, to cud in comfort and to come to the
milking willingly and without excitement. If they keep out of mischief in the process, so much the better.

However idyllic the relationship between the goatherd and his
flock, this pattern is liable to rupture if the goats are physically
uncomfortable. The commonest form of discomfort is occasioned
by the weather.

Here we come upon a curious anomaly. Goats are notorious for
their objection to wind and rain, and for the haste with which they
rush to shelter from the least inclemency; and goats are famed for
their hardy endurance of the windiest and wettest rigours of our
climate on the mountains and cliffs of the Atlantic coast. The feral
goats of the mountain tops are but a generation or two removed

from the hot-house pets of the goat house. Many a winter I have seen our milk-recorded herd of British Saanens grazing a ridge in the face of a bitter north-east sleet shower which has driven the thick-coated hill cattle to shelter, and sent even the mountain sheep to a warm bed in the heather.

If the goatherd wants to evoke from his flock the hardiness which is inherent in all goats, first he must consider the profound difference between the central-heating systems of goats and men. Man derives his heat almost entirely from the oxidation, 'burning', of the carbohydrate reserves of his own body. In cold weather he eats foods rich in digestible carbohydrates (in the fuel to bank his fires), and he insulates himself in warm clothes and houses. The main source of heat for the ruminant is that produced by the bacterial fermentation of fibre in the rumen. The fuel which supplies most of the heat has not been digested by the animal; indeed, it is material that no animal can digest. So the ruminant meets the challenge of cold weather by eating yet more indigestible roughages, and so lowering its intake of digestible carbohydrates.

The cental-heating system of all ruminants is the same in principle, but the goat carries the principle to its extreme. The goat, which can in a day eat twice to three times as much as a sheep of the same size, is naturally designed to exploit pastures too poor to support the sheep; to live on them the goat has to eat to capacity. Consequently it is the fate of the goat to carry about with it for ever a radiant heater with two or three times the heat output of that of the sheep, and to carry that sweltering load both winter and summer. Mercifully, she has a thin skin and only a light covering of hair in summer, which is reinforced by the growth of underfluff in winter. Even so, you may examine the necessary generosity of her sweat ducts in the grain of morocco leather.

Being designed along lines which should provide a constant surplus of heat, the goat is sensitive to chills. The temperature at which her digestive processes work best is a relatively high one, and cold conditions of the rumen favour the activity of agents of disease.

Under domesticated conditions, especially when put out to graze good-quality pastures, the goat is often offered a diet so poor in fibre that is her main source of heat, that her rumen temperature is dangerously low and even a summer shower may chill her perilously. She comes into her shed in the evening so puffed up with cold mush that, unless her udder be laid on an insulated floor, she gets chilled again. To protect her from these risks, her owner builds a fine, cosy shed, which warms up quickly when the goat comes in with a little autumn roughage in her. When she goes out the following morning, with nothing but a cold, wet pudding of concentrates and mashed turnips in her, she feels the cold more intensely. The nice, cosy shed will also ensure that the underfluff in the goat's winter coat never develops fully (some scrupulous goatkeepers have been known to comb it out too); so the winds of winter strike on her bare skin. When a really cold spell arrives, the goatherd puts a suet pudding on the stove for herself and an extra pound of maize in the goat's feeding bucket; so the goat eats even less roughage than usual and ungratefully—but logically—dies of pneumonia.

If it is desired to feed goats a highly concentrated diet, as it well may be where land and labour are scarce, then the goats must be of a suitable type, the housing must be consistently warm and the goats must not be expected to supplement their diet by foraging outside in any but the mildest weather.

When goats are expected to obtain a good proportion of their food by foraging for themselves, it is essential that their shelter never becomes heated much above the temperature of the outside air. In cold weather, especially, it is bad practice to feed concentrates of succulents (roots, kale, etc.) unless the goat has an adequate quantity of fibre inside her. The state of affairs cannot always be tested by eye, as the goat may bulge with kids, gas or water; but it can always be tested by pressing with one's knuckles in the left flank: where fibre feels 'puddingy', kids are absent, gas exerts counter-pressure, and water sloshes. In summer-time an alternative to highly nutritious grass should be provided in the form of hay, if nothing fresh and natural is available in the way of roughage. This will improve the goat's appetite for grass, and enable her to graze in cooler and wetter weather.

In winter, dry goats require very little in the way of nutrients, and can get all they need in an hour or two on heather and a little longer on brushwood, tree bark and brambles; there is no point in keeping them out of their shelter when they want to return to it. But they are excellent weather prophets, and will always make the most of a spell of good weather if free to come and go as they want.

When goats have to be supervised the whole time they are grazing, it is most irritating if they refuse to get on with the job, and alternate between short spells of foraging and long spells of cudding. If they are led back to their shed as soon as they start cudding, they will soon learn to take as much as they can hold in one whack, and so save herding time. Though food consumption under this arrangement is increased, the goats do not make such good use of the food they eat, either in the field or for several hours thereafter. The goat's digestion works best when relatively small quantities of fodder are passing rather rapidly through the rumen; the more nutritious the herbage, the greater the losses if fed in big feeds. It is therefore better to herd the goats out for half an hour in the morning and half an hour in the evening than to herd them for two hours on end. This is lighter on labour and feeding and the goat's digestion, and makes a heavier job for the milker.

Under most conditions, unfortunately, the control of goats requires more than a good command of their mental and physical comfort—it needs fences.

Any fence less than 4 ft (1·25 m) high will be jumped, but over 3 ft 6 in (1·1 m) high it will not be cleared by the milkers, who will bruise and leave portions of their udders on the top of the fence. The only exception to this rule is the electric fence, which goats are understandably loath to jump even at 3 ft 6 in. A standard 4-ft stock fence, with seven wires spaced at 6-in (15 cm) intervals, will hold goats of a reasonably placid disposition so long as the wires remain dead tight. Eventually the goats will force their way through the wires. At any time a sudden urgent incentive will send most goats over the top of a 4-ft fence of this type. In theory it is possible to keep wires taut, but the theory ignores the fact that wire expands and contracts with changes in air temperature; if tight on a hot May afternoon the wires will probably snap in the frost of the following morning, unless they are 6-gauge or heavier.

Wire-netting fences give a great deal of pleasure to goats, who derive exquisite satisfaction from massaging themselves on the netting. The procedure is for the flock queen to throw her full weight against a section of fence between two posts, and to press her flank slowly along the condemned section; each member of the flock follows her, and then she will take another turn; the game goes on for hours every day, and for as many days as the netting lasts. When the section has collapsed and the goats have had their fill of the crops on the other side, they lie down to cud on the ruins, which provide that degree of insulation from the earth which goats crave. The hard, ugly glint that appears in the eyes of some farmers when goats are mentioned is often due to the pleasure that wire netting provides.

If you run two lines of barb along the wire netting it spoils the fun, but usually does some serious damage sooner or later.

With fences of the Rylock and Bulwark type goats adopt a rather similar technique: they provide massage in the first instance; but with fences of this type it is sufficient that the fence be given a slight slant, at which stage the goats accept the convenience of the ladder steps provided by the horizontal wires, and go over the top. There is then the added embarrassment that they cannot get back again to their own ground when occasion demands.

The most generally applicable solution to the fencing problem is a 4-ft (1·25 m) fence of chain link, with posts at 6-ft (1·8 m) centres, and with two or three support wires. The chain link can be either 2-in (5 cm) mesh and $12\frac{1}{2}$ gauge or 3-in (8 cm) mesh and $10\frac{1}{2}$ gauge. It must be strained.

The method of erecting this fence is as follows: after lining up and erecting the fencing posts and putting the support wires up at 4 in (10 cm) and 4 ft 4 in (1·35 m) above average ground level, link up the 25-yd (23 m) sections, in which the fencing is sold, for the required distance, and lay the fencing flat down beside the fence. If the fence passes over any large humps or hollows the wires of the chain link are unlocked over the appropriate sections at both top and bottom and twisted up or down to give the required curve to the bottom line of the fence. Then they are relocked. Now fix one end of the fencing to one corner post and staple a 4-ft (1·25 m) length of 3-in by 2-in (8 × 5 cm) timber across the other end of the fencing at a point which is about 3 yd (2·75 m) plus 1 yd (0·9 m) for every 25 yd of fence length, from the other corner post. Then take a straining wire from the dead centre of this piece of timber to the corner post, taking care that the wire doesn't pass through the fencing. Finally strain this wire, feeding the end round the corner post until the fencing is rigid. Make fast the straining wire, and staple the loose end of the last 3 yd of the fencing to the corner post; should the fencing subsequently slacken, straining on the one wire only will tighten it once more.

An electric fence is satisfactory for temporary fencing, but requires to have three live wires at 12 in, 27 in and 40 in (30, 69 and 102 cm) above ground level; these heights may need adjustment for large or small goats. The goats must be made acquainted with the potentialities of the fence, and learn to respect it, before the fence is charged with the duty of holding them. If they are allowed to break through once, they will accept the sting as the necessary price of liberty and do it again. Male goats sometimes appear to be entirely devoid of a sense of pain, and their control can never be entrusted to an electric fence.

Plate 4.   An electric fence is all that stands between Toggenburg temperament and a field of kale. Note the heights of the three wires.

In more permanent forms of fencing, where a solid top bar is a feature of the design, the height should be increased to 4 ft 6 in (1·4 m) a height which is suitable for all gates into goat enclosures.

In pale or rail fences, the distance between the pales or rails should be under 3 in (8 cm) or between 7 and 8 in (18 and 20 cm) to avoid a goat or kid trapping its head in the space. The wider spacing is not kid-proof. The same limits apply to the distance between rigid bottom rails and ground level.

No type of hedge associates well with goats. A well-kept hedge of *Rhododendron ponticum* bears up best under the relationship; goats accustomed to the sight of it do not eat rhododendron, which is poisonous to them, but a newcomer to the flock may do so; and the accustomed flock will reconnoitre and wriggle a way through if it is possible to do so. A mature beech hedge of 5 ft (1·5 m) or more high will sustain little damage if alternative leafage is available; if grown in the Continental trellis style, such a hedge will be goat-proof as long as it lives; but the normal double-row beech hedge will usually admit a determined goat. Thorn and blackthorn hedges may hold goats as long as they live, and a pure blackthorn hedge will survive many years with goats; but they are rare, and the normal thorn hedge will not stand up to persistent goat browsing. A yew hedge is too dangerous to test. A holly hedge is a goat's paradise. Barberry is also eaten but not avidly. In general, hedges may control passing goat traffic but not persistent attack.

The effectiveness of stone walls as a barrier to goats depends principally on the roughness of the surface. The width of the top is of no significance. Nothing under 4 ft (1·25 m) sheer height is any barrier at all; beyond that height the goat cannot make a clean jump (billies to 5 ft [1·5 m]); but if the surface of the wall is rough, with a few projecting footholds, they may run diagonally up anything up to 6 or 7 ft (1·8 or 2·1 m). As with all goat control, much will depend on the social life of the flock (an elderly flock queen is not given to acrobatics) and on the degree of frustration they experience in their confinement.

While the use of barbed wire in goat fences is generally neither safe nor effective, the addition of one strand of barbed wire at 5 ft (1·5 m) on the 4-ft (1·25 m) chain-link fence makes a combined goat-and-deer-proof fence which is quicker to erect and requires less maintenance than a classic all-wire deer fence. Where goats are kept on a hill farm, such a fence has been held eligible for subidy under the Hill Farming Acts.

In general, it can be said that a fence that is goat-proof is also deer-proof and vice versa. Few deer fences of the 11-wire type are deer-proof or goat-proof; but most of them are a sufficient deterrent for practical purposes.

Many a poor man cannot afford a fence against either deer or goats. The deer carries away a ·22 bullet in its guts, and the goat goes on a tether.

One goat alone on a tether is not a happy sight. A number of goats tethered within sight of each other can form a contented and productive community. Granted that the tethering system has been much abused and that it is very expensive in labour, it provides a perfectly feasible and satisfactory method of goat control under some circumstances. Much of the unqualified objection to tethering is rooted in plain, common snobbery.

The commonest form of tether is the picket tether, where the goat is tied by collar and chain to a stake driven into the ground. The collar is the better for being extra wide; the chain must have swivels at both ends, and be reasonably light. It is helpful if the chain is made of three easily detachable 6-ft (1·8 m) lengths, so that the tether can be lengthened as the day progresses, without excessive fumbling. The chain is attached to a ring which rides and revolves freely in a collar round the top of the stake. The stake must be driven in until the ring is flush with the ground, or the top of the stake will foul the chain. If the local blacksmith is too busy mending tractors, tethering equipment is available from the specialists in goat equipment who advertise in the appropriate papers. Pig tethers are adaptable.

The main snags of this form of tether are that they make for uneconomical use of ground, and can only be used on patches of grass and soft herbage. The goat grazes a ring at the end of its tether and soils the circle around the stake; if you divide a field into rings which do not overlap, you leave a lot of ground ungrazed. As a light chain will snag on any obstacle from a fruiting dock stem upwards, the ground on which it will not snag can be guaranteed devoid of the harder types of herbage in which goat digestion delights. In warm weather, the picket-tethered goat is often enveloped in a thin haze of flies which detract from its peace and content, and from which there is no escape. A spray with insect repellent (rather than an insecticide) helps matters in this respect.

The alternative form of tether, known as the 'running tether', is slightly more cumbrous to establish, but considerably more satisfactory in action. Essentially the running tether consists of a fairly short length of chain—not more than 3 ft (0·9 m) and not less than 18 in (45 cm)—with a swivel, collar and goat at one end, and at the other end a swivel ring which runs along a wire stretched taut along the ground.

The wire may be of any length from 6 or 7 yd (5·5 or 6·4 m) up to several hundred. If of a short length, the ends of the wire will be held by two stakes, the heads of which are driven flush with the ground. Alternatively, a long wire may be lightly strained from the fencing posts across the corner of a field, or right across a narrow field. In this case the goats' chains must be sufficiently short to prevent any possibility of their even considering jumping the fence at each end. A long wire can also be strained between two posts planted in mid-field; but in this case a piece of wood, longer than the length of the tethering chain and goat combined, must be tightly stapled to the wire at both ends to prevent the goat from getting caught up on the post. Such a piece of wood can also be stapled on to the wire at any point along its length to divide it into tethering sections. In this way a dozen or more goats may be tethered along the 'grazing face' of a forage crop; the daily move forward will only call for the removal of the end posts and one strand of wire; the cost of such equipment will be little more than the price of the wire. As against the electric fence, it has the disadvantage that you must spend a few moments attaching each goat to its tethering chain each day, and the advantage that there is no electric circuit to be shorted out by a heavy, wet crop.

The running tether uses a short length of relatively heavy chain which does not snag easily; if it does snag, the goat can apply a great deal more force to release herself than is possible on the picket tether.

The running tether has proved quite satisfactory for strip grazing of kale, with grazing periods of up to four hours. It has proved entirely satisfactory for goats eradicating thistles and docks from new leas for grazing periods of twelve or fourteen hours; it can be used with a measure of watchfulness right through the midst of a strong old growth of brambles and briars.

How much wind and rain the tethered goat can tolerate depends on the kind of fodder she is getting. It is quite proper for the owner of the goat receiving 4 to 5 lb (1·8 to 2·25 kg) of concentrates a day to rush her corn bin back to the goat shed as soon as a rain cloud darkens the face of the summer sun. But sterner diets make goats of sterner stuff. If an elementary shelter is strategically placed at the windward end of a running tether, most goats can be tethered on coarse weeds and grass most days from May to September.

Flies and sun are another problem. As far as the former are concerned, the goat on a running tether is little worse off than the goat with free range over a few low-lying acres. Sun intensity in Britain seldom reaches the goat's limits of endurance, provided the goat has access to water; but in damp, windless heat the goat is best kept in the shade during the worst of the day: it is not the sun that troubles the goat so much as hot humidity.

For any goatkeeper who is not a pedigree breeder of the first class, there is no purpose served in rearing male kids for anything except meat. There is always an ample supply of pedigree males at prices well below the cost of rearing. Readers who care to disregard this excellent advice must accept the consequences and a very troublesome task of control.

From the age of three months, the male kid may be capable of service; outside the normal breeding season it is unlikely that he will give effective service, but the risk exists of spoiling one of the female kids with whom he is running, or all of them for that matter. For this reason males are usually segregated from the age of three months; the old males living in isolation are apt to maltreat male kids, so that each male kid has to be reared by himself and have all his food carried to him and all his social education provided for him by the goatherd. But some older males willingly accept a male kid for company. He must be taught to lead, and to obey, and not be permitted to play. Above all, he must not be teased.

Fig. 4. The Spanish halter
A device used in both Spain and Germany on goats grazing mature orchards and olive groves, to prevent them from reaching up into the branches.

The adult male is dangerous to all children and to any man or woman who is afraid of him: he is a nuisance to anyone who he thinks might be afraid of him. To keep him in manageable condition, he must be handled and exercised regularly and very firmly. The method of reducing an unmanageable billy to subjection has already been detailed.

Fences to hold adult males must be at least 4 ft 6 in (1·4 m) high; if with a solid top bar, they need to be 5ft (1·5 m). Yarding, stalling equipment, gates, etc., must be of the strength appropriate to control cattle or ponies.

Chapter Five

Housing

Over a number of years I have been walking, through wind and rain, over the downs, over the hills of Northumberland and Galloway and among the mountains of Wales and Scotland. In all those places, and in all the wind and rain, I have stopped every hour or so to light my pipe; very seldom have I failed in the attempt. Wherever there is a wealth of ground feature and rock, there is a wealth of nooks and crannies wherein one can light a pipe in a wet gale. In such nooks and crannies the wild goats make their bed. The first duty of a goat house is to provide equivalent shelter; quite a number of goat houses fail to perform it.

The wild goat is a great deal more particular about its dormitory than the deer, or hill cattle or hill sheep. Every dormitory of the wild goat that I have seen, in Galloway, Wales and the Scottish Highlands and islands, has boasted a carpet of a depth and insulating capacity unobtainable even at the Dorchester Hotel. Goats originated the principle of under-floor heating, of which the finest example I have seen was in an abandoned shepherd's cottage, near Back Hill o' the Bush in the Galloway hills, the winter quarters of the feral goats which summered around the Monument on the Newton Stewart road. Here the goats reclined on a 4-ft (1·25 m) deep compost of accumulated droppings.

The goats in the caves of Carradale Island had a bigger area to cover and had not achieved the same depth—but even the summer couches of the goats of Lochbuie in Mull were quite luxuriously lined. The need for luxury in this respect is clearly considerable; the need may explain why the feral goats return each evening to the same sleeping place, and the domesticated goat returns unbidden to its shed. When she has too soft and rich a fill of early-summer herbage, or when snowstorms and torrential rain prevent her from getting an adequate bulk of winter roughage, it must be a real problem to the thin-skinned goat to keep her belly warm at night.

A further aid to the solution of this problem is found in the social nature of the feral goat dormitory. On warm summer nights, each goat has its separate couch; but if conditions are cold, there is much sharing of beds, and only one flank of each goat is exposed to the night air. Undoubtedly many cases of pneumonia, gastro-enteritis and mastitis might never develop if the goat were free to provide herself with a companionable hot-water bottle on the critical evening!

These features of the feral goat dormitory are not necessarily desirable in the goat house because they are natural. But there is little doubt that they are natural because they are desirable.

Our goat houses must inevitably be, to some extent, a compromise between that which is most comfortable and health-giving for the goat and that which is most convenient and economic from the point of view of human management. The design of cow houses has been inspired, most markedly, by the desire to ease the disposal of the dung the cow makes while she is being milked, fed and sheltered. But goat houses are designed for the prime object of enabling the feeding of concentrates.

Under most systems of management, it is economic to feed the goats concentrates at some time of the year. But the appetite of the goat is so elastic that one strong and hungry goat is quite capable of eating three or four times the ration prescribed to her, with ill consequences to her own health and the productivity of her companions. So goats can seldom be trusted to share their concentrates equitably at a trough, like cattle and sheep. They must be isolated from one another when consuming concentrates.

There is a remarkable unanimity about the layout of goat houses as a consequence. The goat house consists, essentially, of a feeding passage at which the goat is either held by the neck in a 'heck', tied in a stall, or penned in a loose box, according to the degree of comfort to which her productivity entitles her.

The heck system requires 7 sq ft (0·65 m²) of floor area per goat, and is the prevalent system where the average daily yield is under 3 pints (1·7 litres) per head (e.g. in Morocco and Pyrenees). It permits the goat a bare minimum of movement and comfort, and is not generally practicable unless the goats are on range for the best part of the day. The fidgety goat cannot produce satisfactorily if unduly restrained.

Plate 5. Inside the Malpas goat house where the modern dairy goat has acquired several of her valuable qualities, one is impressed with the orderliness and careful detail that lie behind the successes bred there. A commercial dairy farmer would prefer fewer shelves on which coliform bacteria could accumulate in dung dust.

The stall system requires 10 to 16 sq ft (0·93 to 1·5 m²) of floor area per goat, and is earned by goats giving up to 6 pints (3·4 litres) a day. Though movement is still restricted, goats can give satisfactory production on this limited liberty even when all their diet is hand-fed. It is, in fact, the commonest system of goat housing in cool goat-dairying districts.

The loose-box system calls for the provision of 20 to 30 sq ft (1·9 to 2·8 m²) per goat, but the yield of the goat must rise to at least a gallon a day before it can justify the cost of the extra liberty and shelter the loose box provides. With the extra liberty and shelter the goats can produce high yields on a hand-fed diet. As it isolates the goats more completely and securely than any other method, the loose box is a very convenient arrangement for feeding concentrates in large quantities, rationed to individual requirements.

Though goats should be isolated while being fed concentrates, there appears to be no great advantage, and some disadvantage, in isolating them at other times. Goat houses which are nothing more than dormitories require no partitioning whatever, but a floor area of about 12 to 15 sq ft (1·1 to 1·4 m²) per goat. Accommodation for controlled feeding of concentrates and uncontrolled feeding of other food may be provided separately or under the same roof. If provided under the same roof, the floor area allowance per goat will be 20 to 30 sq ft (1·9 to 2·8 m²).

No arrangement is cheaper to provide than the heck system; with it the covered feeding yard and dormitory—the communal goat house, to give it a label—cannot compete. In a cool wet climate, such as Britain, it is quite as expensive to provide a communal goat house as it is to provide stall housing: though there is less partitioning to erect, there is a greater floor area to be adequately sheltered and insulated. In drier and warmer climates, such as that of Spain, the communal goat house does not have to provide such a high standard of shelter, and it replaces the stall system. Under the similar Australian conditions, it could do so equally well.

In comparison with the loose-box goat house, the communal goat house offers a high standard of liberty, of social amenity, interest and comfort at lower cost. This type of housing accords well with the methods of control and feeding advocated in this book, but is subject to one serious objection—the amount of bullying it permits.

Bullying does not normally reach serious proportions in a flock with an active social life and a recognized flock queen, unless some of the goats are horned and some hornless. If this is the case, either the horns or the communal goat house must be dispensed with. Also, as newcomers into a flock have to 'fight their way in', and the process takes some time in a large flock, the communal goat house is an unsuitable form where goats are regularly boarded, or there is much buying and selling of adult stock.

The first duty of the goat house is to shelter. Here we come to the standard instruction that it must provide plenty of fresh air but not draughts. The instruction is misleading, because the more fresh air there is a house the less cause there is for draughts, and especially for the cold down-draughts, which are the most dangerous and discomforting. There is very little draught in the hollow on the hill side, but as soon as the warm air of a goat shed comes in contact with cold, thin-skinned roof and walls, a vicious down-draught is set up which strikes on the goats' backs; at the same time, cold air streams into the warm shed through every chink in the building and, being heavier than warm air, promptly seeks the floor where the goats are lying, and swirls about them.

Plate 6. A kid sleeping box—a detail of management which helped the white kid, Malpas Melba, to establish a world record.

There is no point in paying lip-service to fresh air. Either we want it or we do not want it, or we want but a little of it. If we want a really warm shed we do not want fresh air; unless we provide a central-heating system the goats heat the air; if the air is warm it is not fresh. Fresh air is the same temperature inside the goat house and outside it: if that is the temperature you want your goat house to maintain, you want fresh air; otherwise you want as little fresh air as possible, or a very restricted supply.

The peculiar reverence in which twentieth-century Britain holds fresh air must not deter us from thinking about the stuff. Fresh air has a full content of oxygen; it is very unlikely that in the cosiest goat shed the goats will be in the least embarrassed by shortage of oxygen, so that aspect of the matter can be forgotten. Fresh air has a relatively low moisture content, at least in cold weather, and can absorb moisture from the goats' lungs and skin. Stale, warm, moisture-laden air can absorb less. The effect of such a 'stuffy' atmosphere on the goat's lungs is not very clear; but it is probably true to say that a goat accustomed to such an atmosphere is liable to chill and pneumonia, if suddenly exposed over long periods to cold, dry air. The effect of a 'stuffy' atmosphere on the goat's skin is visible and quite well known: the skin exudes oil which renders the coat exceedingly smooth and glossy, and inhibits the development of the woolly underfluff. For this reason the rugging of goats, even in summer-time, is an old trick to produce the show-ring 'bloom'.

Fresh air has a low bacterial content; stale, wet air is rich in bacteria. But, apart from some rare forms of catarrh and a number of common but sub-clinical infections, no important goat disease is airborne. From this point of view the worst that we can say of stale air is that it is more difficult to produce clean milk in it.

In effect, fresh air in her shed is good for a goat who has to spend a substantial proportion of her life outside; stale air is suitable for a goat that is going to spend most of her life indoors. It is perfectly true that fresh air is more challenging and invigorating to the goat's sytem; that a goat that lives and sleeps in the fresh air attains a *joie de vivre* beyond the reach of the other. But such a goat is not necessarily more productive.

In fact those who serve the Mammon of exploitation must serve faithfully. To produce heavy yields over long periods and through the winter, the goat must have a large intake of concentrates and water. From the point of view of heat production the term 'concentrates' must include all foods in which the percentage of starch equivalent is more than half the percentage of dry matter. A glance at the table of Feeding Values (Table 5, p. 169) will show that such 'bulk' foods as kale and mangolds are in fact more concentrated than oats and some of the oil cakes. These 'succulent' foods are more hostile to heat production than the concentrates, because their nutrients are accompanied by so much water as to direct them towards the udder and away from the body reserves; moreover, the water content is cold, and has to be heated up. The more watery roots contain barely sufficient energy value to heat their own water content to blood heat sufficiently quickly to suit the high-speed digestive system of the goat. Such foods, if fed in quantity through the winter, must be fed in a warm shed and therefore in relatively stale air. Organize a good, thick fug in the goat house and keep it there. A daily spell of *brisk* exercise is all that the goats need to know of fresh air in winter-time. In summer their goat-house design must allow for sufficient ventilation to keep temperatures at a reasonably comfortable level; but this is an easy matter to arrange.

In the house of goats that are to forage out through the winter let there be no vestige of fug. Shelter the goats from wind and rain and snow, but let the air they breathe under cover be no warmer than the outside air. A Dutch barn equipped with sleeping benches and a wind-screened feeding rack is the sort of building that will meet these requirements. In such a building there are no down-draughts from cold walls and roof, because roof and walls are no colder than the air. These conditions will not make for high yields, but they will make for economic and healthy lactations.

Unfortunately there are not a great many goats living under conditions where they can get a substantial proportion of their diet from the natural wastes of winter. Most of their food, and the fuel to keep them warm, must be hand-fed and costs money. The warmer the goats are kept, the less of that costly feeding goes up, as it were, in smoke. So there is a demand for warm sheds for foraging goats; in these there must be a regular inflow of fresh air, or the goats will be unfit to forage outside. To produce a satisfactory shed of this kind for a small flock is undoubtedly more expensive than to feed more hay in an open shed. It is never necessary to provide a warm shed, but the main considerations governing the design of these sheds are given here.

Light wooden structures or houses made of corrugated iron or asbestos sheets all tend to develop chinks in old age. Unless very well ventilated, they produce down-draughts in cold weather, when condensation on roof and walls will add to the discomfort of the goats. The area of the house immediately against the walls is virtually uninhabitable in cold weather, and the floor is too draughty to be slept on.

The problem is depicted in Fig. 5. Ridge ventilation of a house of this type is worse than no ventilation at all, for ridge ventilation fails to extract the warm air before the warm air has made contact with the cold roof surfaces and set up down-draughts and condensation. Though extracting some warm air from one side of the house, ridge ventilation accelerates the cold down-draught on the other side. The warmth of the goats on any one side of the house depends on wind direction: they will be quite cosy one night and lying in a freezing down-draught the next. The provision of hopper-type windows, opening inwards, just below eaves level, in addition to ridge ventilation, diminishes down-draught, but does not stop it. No other type of window below eaves level can do more than intensify discomfort.

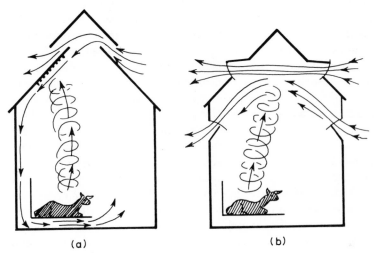

(a)   (b)

Fig. 5. The ventilation of thin-skinned goat houses (a) Ridge ventilation. (b) Rooflight and hopper windows

Roof lights or fixed ventilators on opposite sides of the roof provide the best basic ventilation system. The ventilators on both sides of the roof must always be opened the same amount, unless, perhaps, when a steady gale is blowing from a settled quarter. Ventilators must open outwards. The exact positioning of the ventilators will depend on the pitch of the roof and the ground plan, but generally they should be nearer the eaves than the ridge, in order to extract the warm air before it condenses on the cold roof. Hopper-type windows, below eaves level, will assist this system; but the windows must open inwards, and the eaves must have a deep overhang to prevent driven rain from blowing in on to the goats. Other types of windows are suitable for summer ventilation only.

Apart from a suitable ventilation system, the best hope of warm comfort in a thin-skinned goat house lies in insulation of the walls and roof. Since, even in these energy-conscious days, the British remain reluctant to insulate their own dwellings adequately, it may seem a little ambitious to recommend insulating a goat house. But in fact modern insulating materials are very cheap, readily available and easily handled. In most cases it will cost less to insulate a goat house with material to last its lifetime than to give it one coat of paint. Three main types of material are used: mica flakes, glass-fibre blanket and aluminium foil—all of them are fire-proof, rot-proof, non-absorbent and deterrent to vermin. Very poor meadow hay, loosely packed, is also a good insulator, but lacks the permanent virtue of the others. Mica flakes are laid loose, 1½ in (4 cm) deep, on flat or nearly flat surfaces. They are the cheapest and easiest form of insulation for flat concrete roofs, being covered with paper and sandwiched between two layers of concrete; they operate by trapping a quantity of still air in insulated pockets. Glass-fibre blanket works the same way, but can be draped over or under sloping roof surfaces. It is rather unpleasant to handle: either use leather gloves and close-woven overalls, or bare hands and arms and wash them in cold water when you are finished. Aluminium foil is not a blanket insulator but a heat reflector. The effect it produces is similar to that produced by the other insulators, but it is possibly better than any for retaining heat in a well-ventilated goat house, and for keeping a house cool in hot sunshine. It loses its properties if laid in direct contact with other metals, but is ideal for lining under felt roofs. It is tacked on to the wall or roof surface in overlapping sheets.

Wood itself is a good insulator, and matchboarding is sometimes used for lining corrugated-iron sheds, but it is an expensive form of insulation.

The conversion of old stone- or brick-built buildings is usually the best means of providing warm shelter with a minimum of draughts. It is also usually the least hard on the pocket and the appearance of the countryside. But if new buildings are required, they need not necessarily be ugly. Bales of straw make good building bricks although they are no longer inexpensive; they can be secured in position between two lines of fencing. If set on a dry foundation, thatched on the outside and given a roof with a generous overhang, they will last for fifteen to twenty years. Thatchers can still be found at a price. Six tons of straw will provide walls and gable end for a twenty-goat house.

In districts where straw is scarce, stones tend to be plentiful; for a warm, well-insulated shed there is much to be said for the primitive rough-stone house. It is not beyond the capacity of the average handyman to construct. The wall is made 3 ft (90 cm) wide at the bottom, tapering to 1 ft 6 in (45 cm) wide at the top; the inside of the wall is kept straight and all the taper is on the outside. Cement is used on the inside of the wall, but on the outside only for the last foot at the top. Between the inner and outer stones at the bottom, turfs are rammed as building progresses. At a height of 3 ft, strap stones are laid at intervals across the width of the wall; from that height upwards, the interval between inner and outer stones is filled with small stones and gravel. The outside corners are rounded, and the walls are taken to 6-ft (1·8 m) height or a little more all the way round—with no gable end. The roof, which must be angled or rounded at both ends, may be either felt or thatch. The width of this type of house must not exceed twice the height of the walls, and the roof must be rather steeply pitched. Good models of this type of house are common in the Hebrides.

Concrete-block construction is now relatively cheap, provided the erection is not done by a builder. If there is no objection on the grounds of appearance, cavity blocks make a good goat house. Double-wall construction is preferable, but single wall will do. Flat roofs are the easiest and most satisfactory for concrete-block buildings, but they must be insulated. With flat roofs, two hopper-type windows or ventilators facing each other, opening inwards, with their tops at ceiling level, give the most satisfactory ventilation.

Good advice is more plentiful than hard cash, and inevitably many goats will have to put up with more or less ramshackle houses and a thoroughly perverse system of ventilation. To minimize the discomfort and danger, concentrate attention on the goat's bed. In loose boxes and communal goat houses, the simplest solution is to provide the goat with a sleeping bench. This in its simplest form consists of ⅝-in (16 mm) boarding, 4 ft (1·25 m) long and 2 ft (0·6 m) wide for a single goat (double the width for two), set on legs or propped up on what-have-you, some 9 in to 2 ft (23 to 60 cm) above floor level. This should be set 1 ft (30 cm) or more away from the walls of the house, to avoid wall draughts, and may be improved by the addition of a 1-ft backboard, or by being set against a partition wall. The provision of wings at either end may be helpful in a draughty communal house, but is seldom necessary in a loose box. Where down-draughts are a problem, the sleeping bench can consist of a box, 3 ft 6 in to 4 ft (1·1 to 1·25 m) long, 2 ft to 3 ft (0·6 to 0·9 m) wide and 2 ft to 4 ft (0.6 to 1·25 m) deep set on its side and raised off the floor. On such a sleeping bench, a goat can sleep in an open field a great deal more comfortably than most goats sleep in a warm shed.

The stall system of housing does not lend itself to sleeping benches. You can provide a raised wooden floor to the stall; but it will rot and stink and grow slimy very quickly, even if covered with straw or peat. Alternatively, you can provide a raised slatted floor, a piece of equipment which all goats loathe: if you cover it with straw it will feel less repulsive to the goat's feet and more warming to its belly; but there is a lot of work involved in moving straw and slatted floor sections to clean underneath them.

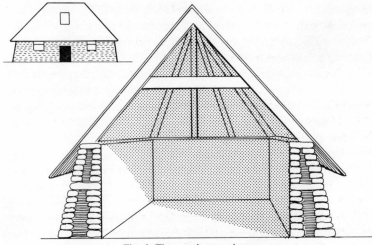

Fig. 6. The rough-stone house

If the house is built with well-insulated walls and roof and is substantially free of draughts, there is no reason why the goats should not sleep on the floor. Many goats have to sleep on floors in a whistling, icy draught. In either case the nature of the floor is important. It should be dry and well insulated. This end is attainable in man's way or in the goat's way. Man's way is to lay down a waterproof concrete floor, sloped to a central drainage point, and to cover the floor thickly with straw. The goat's way is to lay down a hot bed of accumulated droppings, which absorbs and evaporates the moisture in the urine and renders inoffensive the urine's nitrogen content.

The goat's way produces the most comfortable bed and the most sweet-smelling goat house under normally favourable conditions. The goat's way is also most economical in labour and materials. But for its proper functioning the hot-bed system of drainage needs a porous floor. High-yielding milkers on a sappy diet have an intake of 4 to 6 parts water for each 1 part dry matter, and an output of about 10 parts water to 1 part dry matter. This is too much for the absorption powers of the natural hot bed, even on a porous floor. A 2-in (5 cm) layer of peat moss for each 2-in layer of droppings will make a more absorbent bed, and will suffice for normally fed goats.

The hot bed is very rich in bacteria of the coliform type. As the presence of such bacteria in milk is a standard indication of bad dairy hygiene, it follows that clean milk production is rather more difficult where a hot-bed system is used.

Even the stalled goat distributes her manurial largesse over a remarkably wide area, and the quantity of straw needed to keep her dry on a concrete drainage floor is a very serious consideration. Unless this straw can be composted with droppings and liquid manure into a valuable fertilizer for use on the holding, or for sale, the straw is going to be a very heavy charge on the enterprise.

In laying down a concrete drainage floor for a goat house, conservation of liquid manure should aways be arranged. The other main points to consider are: that it is generally inconvenient to drain a pen through the doorway; that goat droppings and straw are a wonderful combination for pipe-blocking, so drainage channels should be open; that one pen or stall should not drain into the next; that the drainage grooves so familiar on cow standings are inappropriate in goat sheds, where the first goat dropping to land in them will frustrate their purpose. The provision of a two-way trap to divert the liquid manure into, and the washing-down water away from, the liquid manure tank, is usually an unnecessary refinement in goat houses. To damp the quantity of straw produced from a concrete-floored goat house, to the state in which it will compost satisfactorily, requires far more water than the goats will produce; so the addition of washing-down water to the liquid manure tank will not go amiss.

The chill that strikes through an uninsulated concrete floor will reach through several inches of straw. If a new floor is being laid it is well worth while to insulate it by one or other of the following methods, neither of which call for much extra expense or labour.

Method 1. Lay the foundations of the drainage floor in the form of heavy shingle or large chippings, no stone of which is smaller than a hen's egg. Lay this material, unmixed, to a depth of about 6 in (15 cm) and form it into the rough design of your drainage floor. On to this bed of loose stone pour a very sloppy mixture of 4 parts sand to 1 part cement, adding to the mixing water recommended quantities of compound known as 'Cheecol'. Allow $1\frac{1}{4}$ to $1\frac{1}{2}$ cwt (63·5 to 76·2 kg) of cement for a ton (1,016 kg) of stone. This will percolate through the bed of stones, lodging only where stones touch one another. It hardens quite quickly, and will give you a honecomb of still air on which to finish your floor; have no fear of its strength, for the system has been successfully used on arterial roads. For a drainage floor it is preferable to finish the surface with a waterproofing compound (e.g. Sealocrete). But Method 1 can be modified to produce a porous floor, in which case you will finish with a mixture of 1 part cement to 3 or 4 parts fine clinker and spread the surface with earth or sand.

Method 2 requires a tractor. Thoroughly till the building site to a depth of 9 in (23 cm). With a post-hole digger, excavate 6 in (15 cm) diameter post holes at 2 ft 6in (75 cm) intervals over the whole site. Fill these with concrete. Then as quickly as possible, and with the minimum of trampling, lay a first layer of concrete straight on to the loosened earth. Finish your floor in the normal way, with a waterproof surface. In due course the earth subsides, leaving the floor supported on concrete pillars.

Such a thing as a standard goat house does not exist. The expenditure of architectural thought on goat houses has been small in quantity, if excellent in quality. It has therefore seemed more helpful to provide basic data and food for thought than specific designs for imaginary, and probably non-existent, conditions. Plans of small goat houses and open-fronted shelters are available from the British Goat Society.

Plate 5 shows an excellent example of the loose-box system of housings, and a number of useful details. Note the insulation.

Figs. 7 and 8 illustrate a practical layout for goat yarding, designed by a goatkeeping architect with many years of practical experience. A yard of this type facilitates the feeding of bulky foods—a most laborious task when goats are fed in stalls or loose boxes. It permits development of the social life of the herd, so necessary to good control, and quite impossible for hand-fed goats kept by other means. It keeps the hand-fed goat in a healthy environment of fresh air and exercise.

The system presented here is an unusual solution to a number of

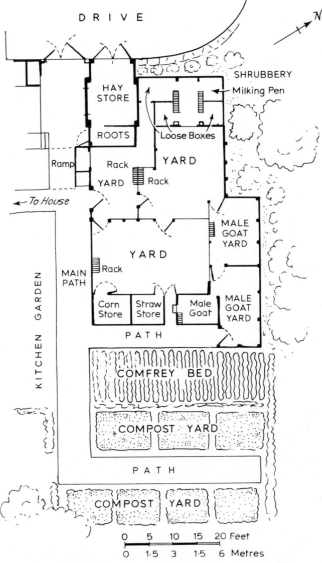

Fig. 7. Sketch plan of practical goat yards, Hanchurch Yews

goatkeeping problems, which has stood the test of time and milk recording. It has every advantage over more popular systems of stall feeding and a great many advantages over the less satisfactory forms of free range—not the least of these advantages being that the whole of the goats' output of dung on a bulky diet is available for composting and application where required. Most of it is deposited in the open air and open spaces of the yards, where it is most easily collected.

Fig. 9 illustrates an American type of goat-feeding rack which is

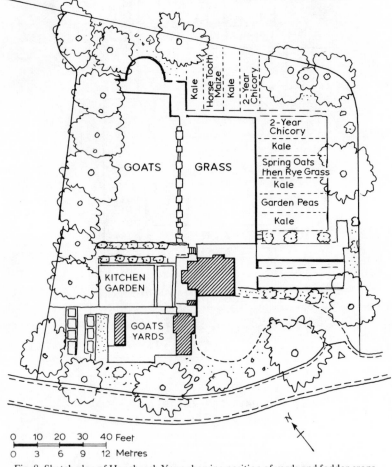

Fig. 8. Sketch plan of Hanchurch Yews showing position of yards and fodder crops

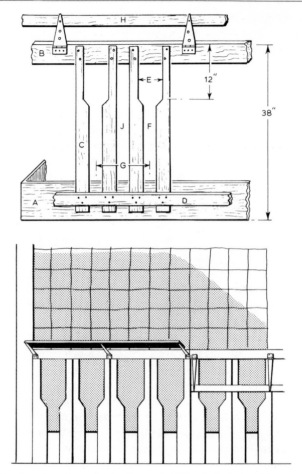

Fig. 9. The American-type goat-feeding rack (After I. A. Richards)

A. Base board of the side of feed box.
B. Top batten of feed box, 38 in (96 cm) from base A.
C. 3 in by 1 in (8 × 2·5 cm) stanchion—cut-out piece, 12 in by 1½ in (30 × 4 cm).
D. Foot rest, needed for goatlings and kids only to enable them to reach the head opening.
E. Head opening: for kids, 6 in (15 cm); for milkers, 7 in (18 cm); for males, 9 in (23 cm).
F. Neck slot: for kids, 3 in; for milkers, 4 in (10 cm); for males, 6 in.
G. Distance between centres: for kids, 12 in; for milkers, 18 in (45 cm); for males, 24 in (60 cm).
H. Batten on hinges, lowered to lock heads in the stanchions.
J. Where J is more than 3 in, it is filled.

usefully versatile. With the locking bar lowered it becomes a free-for-all hay rack, in which the goats can be isolated for concentrate feeding, and held for milking if required. With locking bar raised it becomes a free-for-all hay rack, in which the goat can nose around for the bits she most fancies but has great difficulty in wasting her second choice. The degree of isolation provided with the locking bar lowered is probably insufficient for feeding concentrates in large quantities and individual rations. But it is quite adequate for feeding the standard allowance of 2 lb (0.9 kg) of concentrates per head suggested in this book.

With a 3-ft (90 cm) feeding passage down the centre, this rack could be standard equipment in the communal goat house, the only other equipment needed being sleeping benches and drinking bowls.

If a higher degree of control of concentrate feeding is required, this would be as well provided by a separate milking parlour.

Of the smaller items of goat equipment, the most important is the hay rack. The nature of the ideal goat hay rack is a subject of much controversy. Some people favour hay nets. Certainly they are portable, useful therefore in the show-yard; and, using hay nets, it is possible, as it were, to fill the goats' hay racks in the field where you cut their fodder. But hay nets are a source of much frustration to the goat who cannot get at the bit of fodder she wants: sometimes the goat succeeds in pulling the hay net down and spilling and wasting half the contents; at other times the hay net avenges the attempt by catching the goat by a front leg. They are a source of equal frustration to the attendant who cannot undo the knot at the top, and then the hay net has been known to avenge the insult it earns, by removing the buttons from the attendant's coat. On the whole, hay nets are to be avoided.

The wooden rack in general use is about 2 ft (0·6 m) deep, with wooden bars $1\frac{1}{4}$ in (32 mm) square, set $1\frac{1}{2}$ in (38 mm) apart. The hay slides through them more satisfactorily if the bars are bevelled from $1\frac{1}{4}$ in wide at the outside to 1 in (25 mm) wide on the inside. In making such a rack, screw the bottom of the bars *inside* the bottom rail of the rack, and the top of the bars *outside* the top rail of the rack. In loose boxes it is an economy to place a two-sided rack on the pen partitions, to serve both occupants of the adjacent pens.

Iron hay racks of similar design are obtainable from the goat-equipment specialists.

Another type of hay rack which has enjoyed some popularity is not a rack at all but a box, with a lid on top, fastened to the side the goat cannot reach, and a hole (7 in [18 cm] diameter) in the other side, through which the goat (hornless) puts her head and snuffs around for the bits she wants, without wasting the rest. This device does reduce wastage, but it is doubtful if the goat enjoys eating with her head in a small dark, dusty box.

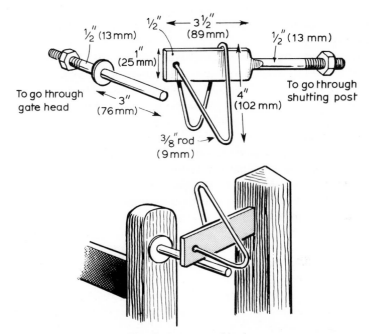

Fig. 10. A goat-proof latch

The same principle of free choice is honoured in the American type of rack, shown in Fig. 9—but here free choice is permitted under lighter and airier conditions, and it is even more troublesome for the goat to waste hay. This rack is also very much handier than any of the other designs mentioned for feeding long, thick-stemmed foods such as kale, hogweed, artichoke tops and tree branches. These foods otherwise have to be chopped and fed in pails or *individually suspended*!

Drinking arrangements in goat houses appear to be rather primitive at first sight. Buckets predominate. They have to be emptied

and refilled whenever a few hay seeds have contaminated the surface, for the goat is a fussy drinker of cold water; but she will
nevertheless drink up to 5 or 6 gallons (22·7 or 27·3 litres) of water
in a day—a lot to carry to one goat. Goats drink quite willingly out
of automatic cow-drinking bowls, and the more old-fashioned,
ever-flowing gutter system suits them well. But goats prefer and
sometimes need to have their water with the chill off; moreover,
such foods as sugar-beet pulp (soaked) come in buckets, and the
miscellany of gruels and skim-milk drinks which figure in the diet of
high yielders also come in buckets—so buckets remain. Nevertheless, for a commercial herd, automatic drinkers and an electrically
heated and thermostat-controlled supply would be cheaper.

Goats are infernally inspired to be able to undo almost any kind
of latch, except the simple bolt, which is an eternal nuisance to the
attendant. As a contribution to goat-house equipment, Fig. 10
illustrates, with more hope than confidence, a latch that is believed
to be goat-proof. This is essentially a large-sized variation on a
common latch which goats do defeat. Perhaps the 3-in (76 mm)
projecting bar might be a little embarrassing in the more compressed loose-box houses, but it is excellent for main doors and
gates, which have to withstand goats.

Chapter Six

The Principles of Goat Feeding

ENERGY REQUIREMENTS

All life needs energy. Animal life needs energy to move, maintain a convenient temperature, reproduce its kind, feed its young and carry out 'running repairs' to its own organism. This energy is derived from the food the animal eats and the air it breathes. The food is the fuel, and the air provides the oxygen necessary to its conversion into heat and energy. Some of the latent energy value of the food is consumed in the process of digestion.

The average amount of energy needed to maintain an animal in normal condition, without improvement or deterioration or unnecessary exertion, has been found for most kinds of farm stock. The earliest experiments were done with cattle, and for some time it was assumed that the requirements of other livestock were the same as those of cattle, reduced or increased in direct proportion to bodyweight. Experimental work with smaller animals demonstrated that this was not true, but that their actual requirements were approximately in proportion not to their weight but to their surface area. More exact measurement still proved that, in computing the 'maintenance-energy requirements' of livestock, we have to take into account not only the larger surface area in proportion to weight that goes with diminishing size, but also the higher metabolic rate—the greater speed of all the vital processes—of smaller animals. On the basis of these latest observations, a formula for maintenance-energy requirements was worked out which is said to be applicable over the whole range of animal size, from whale to mouse.

The unit of food energy used in Britain and much of Europe was the net energy derived from 1 lb (454 g) of starch by a bullock—that is, the latent energy of the pound of starch, less the energy lost in the process of digestion. The energy value of samples of other feeding stuffs has been found by feeding trials, mainly with cattle and sheep, and related to the energy value of starch by a figure called the starch equivalent. Thus 100 lb (45·4 kg) of a sample of good meadow hay fed to wedder sheep was found to have the same energy value as 35 lb (15·9 kg) of starch, and the starch equivalent of good meadow hay is said to be 35. Samples vary, as do the digestive efficiencies of the animals used in the trials, and the accuracy of the scientists' measurements. But these figures provide a useful guide to the energy value of feeding stuffs. Modern methods of calculation, based on metric measurements, are detailed in Appendix 4.

Experiments on the maintenance-energy requirements of the goat have been done in America. The results agree with expectations based on the 'universal' formula already mentioned. The figure is 0·9 lb (411 g) starch equivalent per 100 lb (45·4 kg) liveweight per day.

ENERGY FOR PRODUCTION

We want our goats not only to maintain their condition, but also to produce: mainly milk, but also kids and flesh and, in some cases, hair. To do so, the goat must receive a ration, in addition to her maintenance ration, and according with the kind and quantity of production expected from her.

Though many books on goats published in Britain have given the goats'-milk production ration as $2\frac{1}{2}$ lb (1·1 kg) of starch equivalent (or $3\frac{1}{2}$ lb [1·6 kg] of balanced concentrates) per gallon (4·5 litres) of milk produced—which is precisely the ration recommended for the dairy cow—this ration is inadequate for the goat.

The inadequacy of the $2\frac{1}{2}$ lb (1·1 kg) per gallon (4·5 litres) ration was demonstrated in the course of the basic work on goat nutrition and productivity by Asdell and Turner in the United States. The report of this work was published by the Department of Agriculture for the U.S.A. as long ago as 1938 (*Missouri Agricultural Experimental Station Research Bulletin 291*); copies of the report have been gathering dust on the shelves of British libraries ever since. Of the valuable store of knowledge which the report contained, one fragment emerged—that (mediocre) goats convert their food into milk with greater efficiency than (mediocre) cows. This lent a spurious validity to the 2½ lb per gallon ration. In fact the goat needs fully 3 lb (1·4 kg) of starch equivalent per gallon of milk produced.

The inadequacy of the 2½ lb (1·1 kg) per gallon (4·5 litres) ration is further testified by the empirical advice of leading pedigree breeders and goat dairy farmers. In *Modern Dairy Goats* (1949), by J. and H. Shields, lip-service is paid to the 2½-lb ration, but the maintenance ration there recommended includes hay, greens, silage, etc., in unspecified quantities, and 1 lb (454 g) of concentrates in addition. This would provide 1¼ lb (567 g) more starch equivalent than the largest goat in Britain could conceivably need as a maintenance ration, and would compensate for the deficiency in the production ration of a goat giving up to 2 gallons (9·1 litres). Typical empirical advice, which works well for high-yielding goats, and is very extravagant for lower-yielding goats.

The facts of the case are:
The goat generally produces *more* milk than the cow from the same quantity of nutrients.
The goat uses *less* food for *maintenance* than the cow.
The goat uses *more* food for *digestion* and *metabolism* than the cow.

These facts are just what plain common sense might suggest to us, without any help from the scientist. We have long known that the goat weighing about 100 lb (45·4 kg) produced 5 lb (2·25 kg) of milk a day with very little encouragement; and that the cow weighing about 1,000 lb (454 kg) produced about 20 lb (9·1 kg) of milk a day without exerting herself unduly. As the nutrients in the milk must come from nutrients in the food of the producer, each 100 lb of goat must eat, chew, digest and metabolize* (i.e. turn into milk) about twice as much food a day as each 100 lb of cow. All these processes require energy: 40 gallons (181·8 litres) of blood are pumped through the udder of the cow for each 1 lb (454 g) of milk produced. The faster they are carried out, the more energy is required per lb of food processed—as the motorist would say: 'the

---

* 'Metabolism' is the process of changing digested food into living matter and living matter into excretable material.

higher the m.p.h. the lower the m.p.g.'. So the goat needs more for processing—but as 100 lb of goat gets through its food at twice the speed of 100 lb of cow, it requires less maintenance while disposing of a given quantity.

A total of 3 lb (1·4 kg) of starch equivalent will provide a goat with the energy required for the making of a gallon of milk with 3·7 per cent of butter fat. If the milk contains more or less than 3·7 per cent, it will require more or less energy in the making. Average goats' milk contains about 3·8 per cent of butter fat; few goat-keepers go to the expense of having milk samples regularly tested for butter fat; to be on the safe side, the production ration for a gallon of goats' milk of unknown content should contain 3½ lb (1·6 kg) of starch equivalent.

PROTEIN REQUIREMENT

As well as energy—in the form of starches, fats and sugars—milk, hair and flesh contain a small proportion of nitrogen compounds (protein, etc.) and minerals. For maintenance and production the goat's food must contain a suitable proportion of these elements, in a suitable form.

Animals can digest nitrogen only in the form of complicated organic compounds—true proteins. But the digestive tract of goats and other ruminants is inhabited by bacteria which can utilize simple nitrogen compounds such as ammonia and urea, and convert them into true proteins. Such microbial protein is highly digestible. The normal breakdown of body cells in the wear and tear of living releases into the blood, and thence into the digestive tract, almost as much nitrogen, in the form of ammonia and urea, as the goat needs to repair the wear and tear. But the protein of flesh and milk is built of a number of different nitrogen compounds (amino acids), some of which the bacteria of the digestive tract are unable to construct. So a proportion of true protein is needed in the diet, even for maintenance; more, of course, is needed for growth or milk production. But quite how much is needed it is impossible at present to say with accuracy; for we do not know either the proportion in which the various amino acids are needed to make milk and flesh, nor the proportions in which they are obtainable from the various foods. Furthermore, the bacteria of the digestive tract not only make protein out of ammonia and urea, but also break down the protein in the diet into ammonia and urea, attacking the protein of some foods more than that of others.

Consequently, protein is used most economically if it is offered in moderate quantities and in a variety of different forms. In practice the following quantities are recommended:

For maintenance: 1 part digestible protein to 10 parts starch equivalent.

For growth: 1 part digestible protein to 7 parts starch equivalent.

For milk production: 0·5 lb (227 g) digestible protein per gallon (4·5 litres) of milk.

These quantities will be found adequate if fed in the form of fresh or dry herbage, or in a mixture of protein-rich cereal meals; if fed as oil cake a little more may be needed. The smell of ammonia in the goat shed in the morning, which reflects the amount of surplus nitrogen being excreted, is a useful warning of too much of the wrong kind of protein in the diet. If the milk yield or growth of a goat, being fed at this recommended rate, falls below expectations, it is more effective to change the form and improve the variety of protein foods in the diet, rather than to increase the overall amount.

To summarize the rationing standard for the goat:

For maintenance: 0.9 lb (411 g) starch equivalent per 100 lb (45·4 kg) bodyweight; 0·09 lb (41 g) digestible protein per 100 lb bodyweight.

For production: 3·25 lb (1·5 kg) starch equivalent per gallon (4·5 litres) of milk; 0·5 lb (227 g) digestible protein per gallon of milk.

MINERALS

While scientific standards for rationing the goat's intake of energy foods and protein are mainly useful to beginners and serve only as a rough check for the practices of experienced goatkeepers, problems arising from the goat's mineral needs afflict all manner of goats and goatkeepers, and constitute the principal difficulty of management in most high-yielding herds.

Mineral needs of farm stock vary widely. The relative requirement of each type of stock depends to a slight extent on its size, to a greater extent on the relative magnitude of its digestive organs, and chiefly on the nature of its product.

Small animal systems work at a higher metabolic rate with greater wear and tear, and need more mineral maintenance. The work of digestion involves the use and loss of large quantities of minerals in the digestive juices; animals with a digestive system large in proportion to their body have large mineral needs. Eggs and milk are mineral-rich products which cause their producers to have big mineral needs.

Plate 7.   British Toggenburg kids in the Kinloch Herd nibbling a molehill in search of minerals.

On all counts the goat has an outstanding mineral requirement. A small body with a high metabolic rate; a digestive system occupying at least a third of its body, and producing milk richer in minerals than the cow's, and greater in volume than the sheep's. Feeding adequate in minerals for other stock is liable to be deficient for the goat.

Most goatkeepers become impressed with this fact by bitter experience, or much reading at an early stage in their apprenticeship. Too often they attempt to remedy the trouble by feeding a 'balanced' mineral supplement. Such mixtures usually serve a purpose at a cost out of proportion to their value to the goat. If designed to meet the requirements of dairy cattle under orthodox systems of management, the mixture will not be balanced for the goat. Even mixtures designed for dairy goats may be balanced or not for the individual goat, depending on its feeding and yield. It should be clearly understood that an excess of minerals throws a heavy strain on the goat's kidneys, which are largely responsible for getting rid of the surplus; an excess of any one mineral is liable to make another non-available; a chronic excess of many minerals deranges the working of the vital processes; some of the 'trace' minerals are acute poisons in excess.

No attempt can be made in a book of this size to deal in full with the functions and source of each of the many mineral elements necessary to the goat. The ensuing section is confined to dealing with troubles likely to arise in practice from a deficiency or excess of minerals in practical goat feeding.

Under British climatic conditions no goat, not even a wild goat giving a minimum of milk, can survive for long without access to raw salt. No selection or combination of British herbage can provide enough sodium and chlorine to meet the needs of the goat's outsize stomach and its mammary gland, which secretes $\frac{1}{4}$ oz (7 g) of salt with each gallon of milk—almost 50 per cent more than in cows' milk. The only wild goat herds in Britain which have survived for over a century have access to the sea-shore, to rock salt or to pastures drenched by sea spray in stormy weather. Most of the domesticated goats of Britain are found in the seaboard counties. While domesticated goats in coastal regions may obtain their salt supplies in the same way, the annual salt intake of a good milker is between 15 and 20 lb (6·8 and 9 kg) of salt a year, apart from the sodium and chlorine provided in organic combination in its normal food. This is so far in excess of its need of any other raw minerals that the provision of a supply of salt, unmixed with other minerals, is essential. Salt may be on offer either as a salt-lick block fixed to the wall of the house, as rock salt hung in a wire-netting bag under cover, or as rough salt in a box fixed sufficiently high to be clear of dirt and droppings.

Salt deficiency results in impaired digestion and lack of appetite. A diet of soft herbage combined with a soiled salt supply occasionally precipitates the symptoms. Anaemia resulting from lack of cobalt, or parasitic worm infestation, can lower the goat's ability to retain salt and produce similar symptoms; special attention to salt supply is necessary during convalescence from these troubles.

CALCIUM AND PHOSPHORUS

These elements are the principal components of the skeleton of the goat, and essential in the chemistry of a variety of vital processes. Calcium, for instance, is concerned in blood clotting and in the control of the metabolic rate—that is, the rate at which the vital processes are carried out—and in nervous control.

Phosphorus is needed for the release of muscular energy, for the digestion of oils and fats and for the making of new body cells for growth, replacement or reproduction.

Calcium and phosphorus are deposited in the bone together: if the needs of the body subsequently call for either of the minerals, in quantities which the current diet cannot provide, both minerals are released from the bones together. The two minerals are always associated, yet they are always opposed in the effect which they have upon the body chemistry. A deficiency of calcium produces an opposite effect to a deficiency of phosphorus.

When there is a deficiency of calcium in the blood, the goat will tend to 'overdo it'. She will eat well, yield well and be highly excitable—and then collapse.

A phosphorus deficiency may take many forms, but is always accompanied by a rather dull and apathetic attitude to life.

Calcium is a brake, and phosphorus is an accelerator. In those capacities they act on the thyroid gland, which governs the metabolic rate—yield, appetite, excitability—and the rate at which calcium phosphate is withdrawn from the skeleton to serve the needs of milk production and flesh formation.

Thyroxine, the secretion of the thyroid gland, not only speeds the release of calcium phosphate, but evicts the calcium fraction from the blood stream into the gut, from which it is excreted in the dung. Since the thyroid gland of the goat is half as big again as the cow's, and proportionately active (relative to the size of the goat, of course), goats excrete a relatively large proportion of the calcium in the diet, and must have a particularly high requirement of calcium.

If dietary phosphorus, combined with other factors, stimulates the thyroid unduly, calcium may be evicted from the blood faster than it can be withdrawn from the skeleton. Milk fever and tetany will then result.

If dietary calcium 'brakes' the thyroid unduly, too little calcium phosphate will be withdrawn from the skeleton to provide for the body's needs of phosphorus. Muscular stiffness, apathy, lowered yield and temporary infertility will then result—and the digestive system will be unable to absorb what phosphorus there may be in the diet.

Calcium and phosphorus deficiency disease in goats giving a good yield, or goatlings in good condition, is almost always due to an excess of one or other mineral in the diet. To produce such an excess, the diet must be specialized and rather expensive.

To produce an excess of calcium, there must be an overwhelming preponderance in the diet of legumes (green or dry), sugar-beet pulp, and seaweed meal, either separately or in combination. To produce an excess of phosphorus, the diet must consist largely of cereal products, oil cakes, young grass or young green cereals.

In all natural herbage there is sufficient calcium to balance the phosphorus content, even for the goat. But phosphorus is in chronically short supply in Nature, and for high levels of fertility and productivity must be applied to the ground, organically, in the form of bone meal, sewage sludge or fish manure, or otherwise. Symptoms of calcium deficiency are therefore 'unnatural', very severe and frequently fatal. Phosphorus deficiency in Nature is chronic and its symptoms are mild and temporary—lowered productivity being the principal one.

On exceedingly poor pastures, during their first winter, when the sunshine-generated vitamin D which is necessary for calcium phosphate absorption is scarce, kids may develop rickets, and adults osteomalacia. Housed stock fed poor hay are more liable. Better feeding with a bigger ration of open air, or vitaimin-D supplement, will cure this form of phosphorus deficiency.

Current needs for minerals for milk production do not have to be met by the diet on a day-to-day basis. The goat has a bank of calcium phosphate in its bones, and can live on credit for much of the year; a well-fed, well-bred goat will do so. But the credit must be good and the bank substantial.

The goat's skeleton is laid down during its first year or two of life.

It is well understood that the young animal must be adequately fed to be a successful producer later in life—rickets is rare. It is less well understood that the diet providing excess calcium or excess phosphorus, though it will produce overt symptoms only in extreme cases, is every bit as detrimental to the building of a well-mineralized skeleton as starvation would be.

Plate 8. R. Haslam's herd on kale.

Excess of calcium in the diet of young stock produces, in extreme cases, muscular stiffness in kids and goatlings during the winter. Excess phosphorus seldom produces many symptoms at all except precocity and superb condition—until the goatling kids and dies. But it has been known to give rise to summer-time rickets in goatlings grazing young corn.

The state of the mineral balance in the skeleton bank of the milking goat will be 'in the red' for the first three or four months of lactation; as yield falls it will be recouped gradually from current diet. This is not possible if rationing to yield is practised in the standard, mechanical, dairy-cow way.

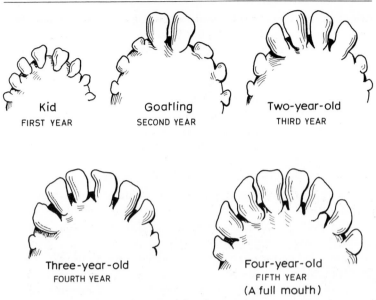

Fig. 11. Determining the age of goats

MAGNESIUM

This in small quantities, is required in the diet, bone and blood of the goat, where it is a necessary companion and assistant to calcium in the chemistry of the body. At least some of the functions of calcium cannot be performed without the presence of magnesium; when the magnesium content of grass—which rises and falls in a seasonal cycle—falls to its lowest levels in May and November, grass-fed goats are liable to tetany. Apart from this trouble, which can be avoided by feeding a bulky alternative to grass or a magnesium supplement at these seasons, lack of magnesium gives rise to little ill health, in practice.

IODINE

The thyroid gland, which is sited in the throat, controls the metabolic rate of goats and other higher animals. It needs a supply of iodine with which to manufacture its secretion—thyroxine.

If the supply of iodine runs short of requirements, the thyroid gland increases in size to make the most of small resources, and a goitre is produced. But the goitre is an unreliable symptom and the least important consequence of iodine deficiency, which can cause a great deal of loss and ill health without any visible or tangible increase in the size of the thyroid gland.

Iodine deficiency produces its own characteristic symptoms: harsh, dry hair; dead, parchment-like skin; still-born (often hairless) kids. Dead female kids born with live male kids are especially symptomatic of a slight but dangerous deficiency—or female kids which die soon after birth, along with male kids that thrive—for the female has a bigger thyroid and iodine need.

Deficient secretion of thyroxine, which is a normal consequence of iodine deficiency, but may be due to other causes too (e.g. poor feeding or a diseased gland), results in lowered vitality and productivity, and poor udder development in first kidders. Since thyroxine is required to release calcium phosphate from the skeleton, deficient secretion may also produce symptoms of phosphorous deficiency.

Recent research has shown that the animal's ability to assimilate vitamin A and carotene depends on its thyroid activity. So iodine deficiency bears in its train the consequences of vitamin-A deficiency as well—retarded growth, infertility, low resistance to infection.

The goat has a thyroid gland which, in proportion to bodyweight, is half as big again as that of the cow. It has a correspondingly higher requirement of iodine. The more productive the goat, the greater her need for iodine.

Because calcium acts as a brake upon the effect of thyroxine in the blood, excess of calcium in the diet calls for additional supplies of iodine to rectify the balance.

The goat's iodine needs are further enhanced if she is fed cabbage or kale which, especially in hot seasons, contain a substance which acts, like calcium, in opposition to thyroxine.

Iodine differs from all the other minerals in that its presence in vegetation is very little related to the species of plant, and almost entirely to the amount of iodine available in the soil. Being highly soluble, iodine is richest in the sea and in districts swept by sea spray: it is rich too on soils that hold their moisture well, peats, clays and humus-rich land. It is easily washed out of thin soils, sands, gravels, limestones; and the more water there is to wash it away—from rainfall or fresh-water flooding—the quicker it goes.

Lime blocks the uptake of iodine from the soil, as it suppresses the effect of thyroxine in the blood. Over-liming produces crops which give rise to an iodine deficiency in the consumer.

Heavy applications of artificial manures and continued cash cropping reduce the iodine content of the soil by lowering its water-holding properties.

Despite the large potential demands for iodine made by the dairy goat, and the varied circumstances in which the soil may be deficient, the actual quantity of iodine required is very small. The application of 1½ cwt (76·2 kg) per acre (0·4 ha) of Chilean nitrate of soda will provide the nitrogenous manuring along with all the iodine needed by goats consuming the crop: 1½ cwt per acre of seaweed meal of fertilizer grade will provide the iodine without inorganic nitrogen.

COPPER

Needed by the goat in minute quantities to aid in the digestion and use of iron in the body. The need is met under most British conditions. In Australia and New Zealand, there are wide areas with an absolute deficiency of copper in the soil. In Britain there are a few notorious districts where the copper in the soil is rendered unavailable to plants by the presence of excess molybdenum—these are the 'teart' and the 'sway-back' areas. But most peatland is rich in molybdenum: if it is heavily limed, the molybdenum becomes available to plants, and the uptake of copper is blocked. Basic slag, which contains molybdenum as well as lime, creates the effect more readily. In the reclamation of peatland this can be a serious consideration, as the Irish have already learnt at some cost.

In the face of available molybdenum it is useless to add copper to the soil, and a copper supplement fed to the goat is the only cure in bad 'teart' districts.

The symptoms of copper deficiency are scours, staring coat, and loss of pigment from the hair, giving the coat a washed-out appearance. The home-made supplement is 1 oz (28 g) copper sulphate dissolved in a pint (0·6 litres) of water, to be poured over 6½ lb (2·9 kg) salt; evaporate the surplus moisture and, when dry, add 1½ lb (680 g) red oxide of iron. Goats may have free access to this mixture.

COBALT

Needed by the ruminant to provide the bacteria of its digestive tract with the raw material from which to synthesize vitamin B_{12}. This vitamin is the antidote to pernicious anaemia; lack of it causes this disease, and encourages acetonaemia—possibly other diseases as well: research workers are busy on the subject. Some, if not all, internal parasites rob their host of this vitamin as it enters the body from the digestive tract.

The sheep has greater need of cobalt than the cow, and the goat appears to have a cobalt requirement four times as great as the sheep's. Since the goat has a daily intake capacity for dry matter at least twice as great as the sheep, the goat would probably be satisfied with a concentration of cobalt in the herbage only twice as great as that which is adequate for the sheep.

There is little consolation in this thought, for the mapping of the areas which are cobalt-deficient for sheep is still in progress, and already includes great stretches of country in the west of Scotland and the south-west counties of England. Moreover, lime depresses the uptake of cobalt, and the lime-rich legumes beloved of goat-keepers require lime-rich soils to grow on. Cobalt is a requirement few goatkeepers can afford to forget.

The proportion of cobalt included in commercial trace-element mixtures has repeatedly been proved worthless for deficient goats. So dissolve 1 oz (28 g) of cobalt sulphate in $\frac{1}{2}$ pint (0·3 litres) of water; wet 6 lb (2·7 kg) of salt with the solution; dry and allow free access. Anaemia consequent on worm infestation, or acetonaemia that accompanies a low-fibre diet, or bitter-tasting milk with a low butter-fat content, can best be treated as a temporary lack of cobalt and countered by adding 1 oz of the above mixture to the concentrates each day for a week. Chronic deficiency due to sheer lack of cobalt is evidenced by a gradual loss of appetite, wasting and sensitivity to cold; as soon as possible the goat grazings should be treated with 2 lb (0·9 kg) cobalt sulphate to the acre; this may be applied mixed with superphosphate or sand. Many fertilizer firms now supply 'cobaltized' superphosphate; or the mixing may be done on the farm in a concrete mixer, or in a barrel, bolted to the jacked-up wheel of a tractor.

Once a deficiency of cobalt is suspected, the diagnosis is easy. Unless there is a dramatic response to the cobalt supplement, prescribed above, within a fortnight, there is no deficiency; unless there is also a heavy worm infestation, the response will occur within three or four days, or not at all.

Cobalt applied to peatland may have an effect similar to that of lime, by releasing latent molybdenum and blocking the uptake of copper. But heather, which thrives on peat, is efficient at concentrating the available cobalt, and goats on heather grazing are less likely to suffer deficiency.

MINERAL SUPPLEMENTS

We may cater for the mineral needs of the goat in three ways: by treatment of the soil on which we grow the goat food; by selection of the species of plant we grow for goat food; and by feeding a concentrated mineral mixture.

The effects we can achieve by treatment of the soil are severely limited. Though most applications are reflected to some extent in the mineral composition of the crop, some minerals are taken up far more readily than others, and the mineral content of the crop is determined more by the typical mineral content of the plant species than by the mineral state of the soil.

Application to the soil of phosphates is generally considered necessary to an economic level of productivity from the land, and is markedly reflected in the phosphate content of most species of plant grown on treated soil. Soil treatment is also the most effective and least costly method of curing iodine deficiencies, except perhaps on chalk soils.

Cobalt and magnesium (in the form of magnesian lime) can be applied to the soil to treat a specific lack. In some cases of cobalt deficiency, and in all cases of magnesium deficiency, the selection of suitable crop plants will have as good effect, or better.

Soil treatment is unlikely to cure animal deficiencies in lime, sodium, chlorine, potassium or copper.

It is not intended to offer an opinion on the relative merits of organic and inorganic manures. But, except in so far as methods which maintain the humus content assist the retention of all the available minerals—especially iodine—the normal run of balanced artificial fertilizers have an effect on the mineral content of the crop which in no significant way differs from that produced by applications of farmyard manure. Farmyard manure increases the copper content of crops above normal levels, but only because most animal feeding already contains more copper than required, and dung is rich in the surplus. Basic slag, however, is not a balanced artificial: it has a remarkable effect in blocking the uptake of trace elements by most crop species, and should be used most warily on land where trace-element deficiencies are suspected. Whatever method of improving land fertility is used—organic or inorganic—if effective, it will raise the available calcium content of the soil, and so lower the content of cobalt in the crops. This fact constitutes a recommendation to feed goats, at least partly, on the produce of uncultivated land. This recommendation seems more relevant than any consideration of rival merits of organic and inorganic manuring—if the merits are, in fact, rival.

To say that treatment of the soil cannot provide a solution to the mineral needs of the goat is to state a happy fact in a distorted and depressing way. The happy fact is that the mineral needs of the goat can be met under an immense range of soil conditions; goats, domesticated and wild, can thrive over a far wider territory than either sheep or cattle.

Sheep and cattle are tied to the more concentrated nutrients of grass by their relatively low food-intake capacity. The large appetite of the goat frees it from this bondage, and permits use of all manner of plants other than grass, even those which have a lower concentration of nutrients. The grasses are, minerally speaking, the most poverty-stricken family in the vegetable world.

To the traditional farmer, pasture is grass. During the past fifty years he has conceded a place to clover as a low-cost grass fertilizer, and a supplier of minerals to more efficient cows. During the past twenty years he has learnt to use lucerne for high yielders, and has heard mention of 'herbs' in the pasture, as an aid to cows whose mineral needs approximate ever more closely to those of the goat as their productive efficiency increases.

'Herbs' is a most unfortunate term, which botanically means everything and nothing. In 'herbs' our great-aunt dabbled benignly, and nasty old village spinsters dabbled wickedly. Herbs have a proper place in the kitchen garden, soups and duck stuffing, but not—the farmer feels—in stock rations and field crops.

For centuries we have been feeding our sheep and cattle on a mixture of grass, legumes, and miscellaneous fodder plants, including weeds such as daisy, buttercup and nettle, acceptable meadow species like plantain and yarrow, and cultivated pasture plants like chicory and burnet. During all of that time every farmer who wasn't stone blind knew that his stock ate most of them, and liked them. Until relatively recently he maintained these species on his fields by sowing out with barn sweepings. Since the introduction of pedigree seeds mixtures, the average annual hay crop has shown no significant increase, the sale of mineral mixtures for stock feeding has risen from near zero to over 40,000 tons (40,642 tonnes) per annum, and mineral-deficiency disease has become a major farm problem. The effect of replacing these miscellaneous pasture plants by grass and clover is to reduce the mineral content of the sward by approximately 20 per cent.

Clovers held a place in the pasture no more honourable than that of the plantain until it became known that clover increased the fertility of the soil. Subsequent study of the legumes proved them to have a high mineral content, and goatkeepers have come to regard clover and its fellow legumes as a necessary goat food. In reality, the common plantain is not only richer in minerals than any legume, but its mineral content is better digested by stock; as a crop plant, plantain can produce more carbohydrates and protein per acre (0·4 ha) than clover. Lucerne yields a good bulk of the major nutrients and is rich in calcium and phosphorus, but there is not one weed of the fields which cannot put its cobalt content to shame.

In pinning our faith to legumes we make the chronic twentieth-century error of blinding ourselves with scraps of science. No one species of plant is good at everything, or there would not be such a variety of plants in existence.

The legumes are expert at fixing atmospheric nitrogen as plant food; but they are most demanding and inefficient in other ways. They must have lime and phosphorus and humus in plenty at their root tips to exist at all; much of the carbohydrate that they fix goes to feed their nodule bacteria, and their content of the major nutrients is either small in comparison with the space they occupy or accompanied by a high percentage of fibre.

Some grass or other can be relied on to cover almost any patch of bare soil with a protective mat and a modicum of organic matter. But grasses are very inefficient at gathering minerals.

Of the other fodder plants of pasture, some, like the plantain and nettle, are miraculously efficient at foraging for minerals near the surface of humus, especially bare humus. Others, like the dandelion and chicory, excel at salvaging from the subsoil minerals that have been washed beyond the horizon of shallow-rooted plants. Such contemptible species as buttercup and yellow rattle keep minerals in circulation on the surface of waterlogged pastures. On hard, acid ground sheep sorrel hunts up lime. In the natural pasture, each of these species in its own way delivers minerals at the feet of the nitrogen-fixing clovers, and at the feet of the grasses which form the protective mat. Such is the natural economy of pasture.

From the foregoing paragraph the goatkeeper is intended to garner two notions—that the other fodder plants of pasture (the 'weeds') are by their nature, and in general, superior in mineral content to both grasses and clovers. This will be easily acceptable by experienced goatkeepers. That the presence of weeds in a pasture increases its general productivity and well-being is a notion that smacks of nonsense. But it is true, and can be seen to be true if we get over the inevitable misunderstanding.

With time and use, a pasture grows weedy and its productivity decreases; so we plough it up, put the land through a rotation of crops and manuring and reseed it to a more productive pasture. The increase in weed content is a certain symptom of lowered productivity; but it is not the cause. The cause is our failure to replace the plant foods and minerals which our use of the pasture and leaching effect of rain remove. In the weakened turf there is room and need for increasing numbers of weeds to provide the deficiencies. But if, after ploughing and putting through a rotation, we reseed with grass, clover and 'other fodder plants' suitable to the soil, the productivity of the sward will be higher than if we omit the 'other fodder plants', and deterioration will be slower.

No special inspiration or insight into the workings of nature is needed to reach these conclusions. Brynmor Thomas and fellow research workers at the Durham University School of Agriculture have investigated accurately the earlier suggestions of R. H. Elliott and Stapledon. Opposite are the facts concerning one of their trial fields at Cockle Park.

It will be noted that the statement that the removal of 'weeds' from our pastures 'lowered their mineral content by 20 per cent' is somewhat charitable to the agricultural advisers responsible.

There is a difference between weed and 'other fodder plants' in that though weeds are all mineral-efficient, some weeds are not fodder for any beast, and their presence in the pasture reduces productivity accordingly. A list of some fodder plants of pasture, with their content of major nutrients and minerals, is to be found in Appendix 2. The intelligent use of this list will provide all the mineral supplements needed in most districts, for goatkeepers who grow most of their own goat food—with the probable exception of salt.

Table 1: The percentage composition of the herbage from swards containing varying percentages of other fodder plants

Constituent	Standard grass-and-clover mixture	With 10 per cent other fodder plants*	With 50 per cent other fodder plants	With 100 per cent other fodder plants
Crude protein	16·75	17·19	16·90	16·94
Crude fibre	21·32	19·49	18·61	15·16
Total ash (silica free)	10·18	11·07	13·01	14·83
Calcium	1·15	1·36	1·60	2·16
Phosphorus	0·29	0·36	0·38	0·41
Magnesium (MgO)	0·42	0·45	0·48	0·52
Sodium (Na_2O)	0·08	0·11	0·14	0·18
Chlorine	0·26	0·30	0·37	0·48

* Later experiments suggest that 7 per cent of other fodder plants produces the optimum yield of major nutrients.

Some of the fodder plants, such as chicory and plantain, can be grown economically as row crops—but in feeding, remember that plantain is laxative. Nettles can be grown as a rotational row crop on a garden scale if the roots are set in bottomless buckets or oil drums sunk in the soil; they are best used for hay. A permanent field crop of nettles for silage and hay is a long overdue agricultural experiment. In general the fodder plants are best sown where they belong—broadcast in a composite pasture. To maintain the pasture's content of fodder plants, it is preferable that the field be grazed by cattle or horses as well as by the selective goat.

The provision of fodder-plant hay and pasture is particularly imperative for the owners of high-yielding goats which show a liability to milk tetany in April-May, and to miscellaneous breeding and metabolism troubles in the autumn. The autumn troubles are due to high-calcium phosphorus deficiency, brought about by the introduction of legume or orthodox meadow hay into the diet—hay making increases the proportion of calcium to phosphorus. Tetany is due to magnesium deficiency, and the following table is self-explanatory:

Table 2: Magnesium content of swards with and without other fodder plants at various stages of the grazing season

	May	*June*	*July*	*Aug.*	*Sept.*
Grass and clover only	0·35	0·39	0·46	0·47	0·45
With 10 per cent other fodder plants	0·39	0·43	0·51	0·51	0·47
With 50 per cent other fodder plants	0·47	0·46	0·52	0·53	0·51
With 100 per cent other fodder plants	0·53	0·48	0·61	0·53	0·54

The owners of not-so-high-yielding goats on thin, permeable soils are, in practice, less bothered with mineral-deficiency troubles when they grow most of their own goat food, because they are incapable of providing an excess of any mineral in the pasture. Only at great expense can they raise the mineral status of the soil to a level at which weeds can be effectively excluded from the sward. Well-made weedy hay from poor meadows has a better balance of minerals than seeds hay from fertile ground.

No one cares to test new notions on their valued stock. The suggestions of this section are inevitably rather novel because the scientific work on which they are founded is recent, and dairy-cattle farmers have not such big mineral needs to cater for as goatkeepers have. But the basic idea that the 'herbs' or other fodder plants have an important place in the pasture and diet of the dairy animal is older than any of us, and has received practical confirmation in recent times at the hands of the Clifton Park system of farming and more recently still, at the hands of the 'organic' school. All that is new here is the exact guidance to the value of individual weeds, and the scientific backing to empirical methods.

Before passing to the alternative method of supplementing mineral requirements by feeding concentrated mineral mixtures, it must be emphasized that the more we investigate mineral-deficiency troubles in goats, the more we find them due to excess of another mineral rather than absolute deficiency. Intelligent as goats are, their brilliance does not reach to the ability to make, with one sniff and lick, a chemical analysis of a complicated mineral mixture, and to ration themselves accordingly. That animal instinct in mineral feeding is highly erratic has been proved in this country and others on a number of occasions. Human skill in mineral feeding is often limited by lack of knowledge of the mixture fed, and of the needs of the stock who receive it.

Mineral mixtures will only be required when high levels of feeding are practised and the diet is unbalanced in minerals, or on the poorest pastures. The standard close-ration mineral mixture is suitable for low levels of nutrition. On high levels of feeding, the mineral balance is usually delicate and the symptoms of deficiency lack clarity. Provided the goatkeeper knows the approximate composition of the mixture he is feeding, and has gathered some understanding of the mechanism of mineral balance in the goat, he has a fair chance of juggling successfully with a selection of mixtures. His chances of success are clouded by feeding any considerable quantity of mineralized compound cake, which introduces a big unknown factor into the reckoning. The best that can be done here is to consider the broad categories of trouble.

(1) *The high-calcium diet* consists almost entirely of high-quality fodder crops—clover, lucerne, oats and tares, kale, comfrey—with perhaps a more or less generous allowance of sugar-beet pulp and 2 or 3 lb (0·9 or 1·4 kg) of concentrates. The danger here is twofold. The wealth of calcium may block the uptake of phosphorus from the diet; excess calcium in the blood may 'brake' the effect of thyroxine, preventing the release of skeletal phosphorus, and putting a strain on resources of iodine. Symptoms are most likely to arise with the introduction of winter-feeding legume hay, etc., and take the form of lowered production and miscellaneous breeding troubles.

If the deficiency is a dual one of iodine and phosphorus there will be a response to seaweed meal, which is rich in iodine; yields will improve with this treatment, but as seaweed meal is poor in phosphorus and very rich in lime, the phosphorus deficiency will become more acute and breeding troubles more accentuated. For this reason seaweed meal should never be used in conjunction with a diet of this type.

A standard close-ratio iodized dairy mixture will meet requirements. A cheaper home-made substitute is 10 lb (4·5 kg) salt, 5 lb (2·25 kg) steamed bone flour, with $\frac{1}{2}$ oz (14 g) potassium iodide dissolved in a little water and sprayed over the dry minerals spread in a tray. Some goats refuse bone-meal mixtures; for these faddists a teaspoon of calcium-phosphate syrup (strictly speaking, calcium lacto-phosphate) a day, and a tablespoon of stock iodide solution ($\frac{1}{2}$ oz potassium iodide in 3 pints [1·7 litres] water), mixed with food or drinking water or administered by dosing.

(2) *The high-phosphate diet* may consist of more than 2 lb (0·9 kg) per day of cereal and oil cake in a ration which either does not exceed 6 lb (2·7 kg) of dry matter altogether, or does not include at least 3 lb (1·4 kg) of legume hay or its equivalent in green legumes. The diet is often met with under circumstances where bulky food of good quality is scarce; the symptoms are precipitated most often in early spring, when the goat is being brought into milk on inadequate roughages; milk fever and tetany are serious consequences; nervous excitement and blood in the milk are warnings.

Goats so fed have a great appetite for seaweed meal in many cases; allowed free access they may consume so much that a phosphorus deficiency results. They should be limited to 6 oz (170 g) of seaweed meal per day, which is as good a supplement as any for a diet of this type. Standard dairy salts are of little use.

Calcium-phosphate syrup and a vitamin-D supplement can be administered to goats who do not take well to seaweed meal. The mixture is a valuable one, because it is well digested and can swamp excess of either calcium or phosphorus in the blood by its correct proportions. But the inclusion of vitamin D constitutes a brake upon the effect of thyroxine and may give rise to iodine deficiency. The use of calcium-phosphate–vitamin-D mixture on goats in late pregnancy has been associated with symptoms of pregnancy tox-aemia—abortion and muscular collapse.

Let no newcomer to goats be in the least alarmed by these complications in feeding mineral mixtures. Correctly fed goats do not need them at all; very high yields can be maintained without them; even over-fat condition along with high yields is possible without indulging in this jugglery. The mineral mixtures are only necessary where a goatkeeper who cannot grow or obtain suitable bulk foods wishes to enter the ranks of pedigree breeders and showmen. For moderate yields, good meadow hays and ordinary grazing, or even hill grazing, are quite adequate. For the highest yields without mineral mixtures it is necessary to feed the goat mainly on high-quality bulk foods incorporating a balanced variety of broad-leaved fodder plants, and not only legumes. Chicory appears to be the only plant which can by itself meet all the goat's needs of major and minor nutrients and minerals and produce an economic bulk of fodder from cultivated land.

VITAMINS

No subject better illustrates the change of balance which is taking place in the science of animal husbandry than a study of vitamins. The ever-deepening knowledge of the normal and natural chemistry of the animal body is continually adding to the list of essential nutrients new 'vitamins', substances whose presence in small amounts is essential to health. The identification of each new vitamin replaces a chemical drug, a surgeon's knife or a pathological frustration by a natural foodstuff.

It is normal practice to classify the vitamins into the fat-soluble (A, D, E, and K) and water-soluble (all the rest) groups. This serves no useful purpose to the farmer, and is a fruitful source of popular error, as in the inaccurate advertisements of vitamins A and D as synonymous 'sunshine vitamins'.

Vitamin A

This has nothing to do with sunshine, and is a colourless substance whose importance to animal life lies mainly in fortifying the body's outer defences of skin and mucous membrane against infection and keeping them in good condition. Its deficiency gives rise to disease of the mucous membranes, particularly that of the eyes, causing night-blindness and soreness, and to a higher susceptibility to infection. It is equally important to man and goat.

The natural sources of the vitamin for the goat are the carotenoid pigments associated with the green colouring matter of plant leaves and the yellow colouring matter of carrots, roots and yellow maize. Carotenoids are absorbed from the digestive tract, and converted into vitamin A in the animal body. The conversion of carotene into vitamin A depends on the activity of the thyroid gland; goats have a thyroid gland half as big again, in proportion to their size, as that of a cow, and they are exceptionally efficient in converting carotenoids into vitamin A.

During the summer months, when green herbage rich in carotene abounds, both goat and cow store a reserve in the liver. In winter, when natural supplies of carotene are scarce, part of these reserves are expended in current use. But a certain proportion remains in the liver to be released, when the goat kids, into the colostrum and first milk.

While winter supplies of carotene are poor, a maintenance allowance is to be found in well-cured hay and silage, while kale and carrots in the diet will help to maintain the vitamin quality of the milk. Seaweed meal made from fresh Ascophyllum has a useful carotene content, and yellow maize is the richest source among concentrates. Some attention to the goat's carotene supplies in winter is well rewarded, as they will digest it more efficiently than the cow, and pass it on in the milk when the human consumer thereof is in most need. If available food will not supply it, a vitamin supplement, such as that sold by Boots and Cooper MacDougall, will serve equally well. Cod-liver oil is not recommended as a source of vitamin A for goats in normal circumstances.

Vitamin-A deficiency in goats should not arise under normally good management. But goats are subject to infestation with two kinds of internal parasites that may affect their vitamin-A reserve. Liver fluke is common among goats sharing waterlogged pastures with sheep and cattle, and where sheep or cattle are subject to antifluke dosing, goats should be included in the routine—the standard sheep dose of hexachlorethane being suitable. Such treatment should be followed up with a vitamin-A supplement, preferably of the dry type. Coccidiosis produces violent symptoms in goats on close-cropped pastures. Infection is confined to warm, damp weather and is commonest in early summer. Coccidia block the uptake of carotenoids from the intestine, and the use of a vitamin-A supplement speeds convalescence.

Vitamin B

This is a complex composed of a number of substances soluble in water, which are required in varying proportions by different animals. Men, pigs and poultry, suffer most from deficiencies. The goat, along with other ruminants, is blessed with bacteria that synthesize most of these vitamins from constituents of foodstuffs in the digestive tract, and is largely independent of a dietary supply of them. However, the process may be thrown out of gear by several possible misfortunes. There may be a shortage of the necessary constituents in the goat's food—a lack of cobalt, for instance, may prevent the synthesis of the vitamin—some cereals and pulses have this effect. Certain types of diet, particularly diets low in fibre, may discourage the proliferation of the bacteria responsible for vitamin synthesis. Parasitic worms may grab the vitamins after the bacteria have synthesized them, and before they are absorbed into the body of the animal. Apart from dangers arising from lack of cobalt, which have already been dealt with, the vigorous persistency with which the bacteria carry out their task will overcome any handicaps to the process which are likely to afflict a reasonably well-tended goat. Nevertheless, a goat reduced to poor and shabby condition by persistent worm infestations or prolonged digestive troubles will benefit from access to a vitamin-B supplement. This may be provided in the form of bran mash, wheat-germ meal, yeast, yeast extract, or proprietary vitamin concentrates. Seaweed meal is a useful medium-level source of supply suitable for regular use to prevent deficiency from arising.

Vitamin C
This is not required in the diet of ruminants, though it is usually present in large quantities. The goat makes vitamin C out of the constituents of the blood. No deficiency of vitamin C in goats has ever been reported, and only rare cases occur among cattle.

Vitamin D
This is necessary for the absorption into the body of calcium and phosphorus, and for their deposition in the bones. Along with calcium, vitamin D acts as a check on the action of thyroxine, and as a control of the metabolic rate and the rate of demineralization of the skeleton during lactation. The goat's requirement of vitamin D is not fixed, but varies with its feeding and phosphorus intake. A goat on a low level of nutrition will build a sound skeleton and keep free of milk fever, tetany and osteomalacia under a very much smaller intake of vitamin D than that needed by a well-bred, well-fed goat. In breeding for high milk yields we automatically breed for higher thyroid activity and larger vitamin-D requirement.

The main source of vitamin D for goats is the action of sunlight on a substance in the skin. Even during winter-time, normal exposure to sky-shine (not even wintry sun) will provide more vitamin D than can be found in the natural diet. For goats on a moderate diet with a moderate yield, we need look no further than the open air for vitamin-D supplies. The high-yielding goat, especially in winter-time needs special attention.

Vitamin D is found in herbage in small quantities, being formed by the action of sunlight on plant substances. The concentration is not great, but the process of hay-making increases it, and theoretically 4 lb (1·8 kg) a day of good hay, preferably clover hay, should provide ample vitamin-D supplies to any goat. In practice it often fails to do so, as instanced by tetany and milk fever. For hay also contains a substance that counteracts the effect of vitamin D. For well-fed goats, be they growing or milking, a winter-time supplement of vitamin D is advisable as a safety measure—and with very high yielders, such a supplement may be continued into the summer with advantage. Cod-liver oil is not advised for a variety of reasons; it is indigestible, unpalatable, and even a small quantity causes big fluctuations in milk yield and butter fat. Combined with typical winter-feeding rations, cod-liver oil is capable of producing vitamin-E deficiency symptoms (muscular degeneration) in all ruminants; in goats it appears to be the only method of inducing a deficiency of vitamin E.

Dry vitamin-D concentrates are obtainable, mixed with other dry vitamins, or on their own. Seaweed meal has a moderate vitamin-D content, and is suitable for regular use mixed in the feed of high-yielding goats at the rate of about 1 oz (28 g) per lb (454 g) of concentrates.

VITAMIN E

This is somewhat enigmatic in its functions: a deficiency causes infertility in rats, fatty degeneration of the muscles of calves, heart failure in cows and muscular degeneration in sheep and goats. Deficiency is easy to produce in rats, is a troublesome disease of calves and can be produced in full-grown ruminants only with great difficulty and through the use of cod-liver oil. For the practical goatkeeper is has interest in only one connection. Where goats are used for calf rearing the readiness with which a dietary supply of vitamin E is passed by the goat into its milk can save the calves from the need for vitamin-E supplement. The dietary supply to the goat may take the form of bran or wheat-germ meal. Such calves should have a dry vitamin-D supplement and not cod-liver oil. The trace element selenium appears to bear the same relation to vitamin E as cobalt does to vitamin B_{12}. Selenium supplements may replace vitamin-E supplements.

VITAMIN K
This is concerned with the clotting of the blood. All ruminants are adequately supplied on farm diets; a deficiency is rarely reported among cattle; a deficiency in goats would be a collector's piece.

APPETITE

Until we know what weight or bulk of food a goat will eat in the day, we cannot tell in what form the pounds of starch equivalent and digestible protein of the scientific ration should be offered. Without some kind of standard for the goat's appetite, no scientific rationing system can be practised.

In fact, no such standard exists. No scientific rationing of goats has ever been practised; every experienced goatkeeper has his own empirical methods of feeding; the literature of the subject is rich in whimsy, quackery and vagueness.

When a scientific worker, inexperienced in goat feeding, is called upon to feed goats in the course of his experiments, he looks for the vital information regarding the goat's appetite and nutritional needs in the international reference work, Morrison's *Feeds & Feeding* (or in Hansson's *Animal Feeding* if he is a Spaniard or Scandinavian). Morrison's *Feeds & Feeding* devotes half of one page to the goat in a 1,200-page book. The sample ration given (presumably for an average goat of about 100 lb [45·4 kg] weight) allows the goat $3\frac{1}{2}$ lb (1·6 kg) of dry matter a day. Hansson allows the goat 2 lb (0·9 kg) dry matter per 100 lb weight each day, as a minimum. A leading British scientist, who used the more generous Morrison formula, writes to me from one of our research institutes as follows: 'When we first started keeping goats we had no provision for isolating them from one another in separate stalls and occasionally a goat would break away from its chain and eat the meal of another goat after having finished its own. This meal being dry caused an acute impactation of the rumen. Such impactation caused the death of three goats when we first started keeping them. . . . This happened twice again but we saved these latter two animals by giving a spoonful of kerosene. . . .' A ration which necessitates goats being kept stricly isolated from one another is unsatisfactory!

Writers on goatkeeping produce a fair number of sample diets; if you happen to have at hand the particular combination of foodstuffs recommended, such diets are useful; if not, not. But at least one can calculate from them the quantity of dry matter per day recommended by practical goatkeepers. The results are illuminating:

Table 3: Quantity of dry matter per day recommended by practical goatkeepers

From	
France (Crépin's *La Chèvre*)	7 lb (3·2 kg) per 100 lb (45·4 kg) weight
Italy (Giulani on Maltese goats)	5 lb (2·25 kg) per 100 lb weight
Germany (Uhlrich)	over 5 lb per 100 lb weight
Britain (*B.G.S. Year Book*):	
Minimum	5 lb per 100 lb weight
Gallon milker	6·5 lb (2·9 kg) per 100 lb weight

In the American experiments on goat nutrition mentioned above, the goats were fed ad lib, and no attempt was made to assess their food capacity as such. But their food intake (of a rather concentrated and monotonous diet) was recorded, and so were their body-weights. Thanks to Professor Asdell, who provided me with unpublished information regarding the diet on which the goats were fed, the following figures can be given:

Seven goats over six months consumed on the average 5·15 lb (2·3 kg) dry matter per 100 lb (45·4 kg) bodyweight per day.

Seven goats during the week of minimum food consumption ate on the average 4·7 lb (2·1 kg) dry matter per 100 lb bodyweight per day.

Seven goats during the week of maximum food consumption ate on the average 6·3 lb (2·9 kg) dry matter per 100 lb bodyweight per day.

The hungriest goat during the week of maximum consumption ate 8 lb (3·6 kg) dry matter per 100 lb bodyweight per day.

Further confirmation of the exceptionally large appetite of the goat can be obtained merely by glancing at the yields of officially recorded goats and relating them to the efficiency with which goats have been shown to convert food to milk. A gallon (4·5 litres) of milk represents between 3·3 lb and 4·5 lb (1·5 and 2 kg) of *starch equivalent* in food intake—that is from 6 to 9 lb (2·7 to 4·1 kg) of normal dry feeding—and many goats give between 1½ and 2 gallons (6·8 and 9·1 litres) of milk a day for long periods.

It seems fair to state that the minimum appetite of the goat is about 5 lb (2·25 kg) of dry matter per 100 lb (45·4 kg) of body-weight, and that well-bred, productive goats can extend their intake capacity to over 8 per cent of bodyweight.

The economic importance of exploiting the goat's appetite to the full is obvious. Even a high-yielding goat can obtain almost the whole of her food requirements from natural herbage during a great part of the year. To illustrate this, the following Table 4 shows how a goat, giving a gallon (4·5 litres) of milk a day, can meet her minimum requirements of phosphorus on different types of grazing. Phosphorus is a good indicator of the presence of other essential nutrients, and grazing which provides a goat in full milk with enough phosphorus will usually supply all the food she needs.

Table 4: Percentage of phosphorus needed in food on a dry-matter basis

If, for every 100 lb (45·4 kg) of bodyweight, the goat eats each day, of dry matter:	The percentage of phosphorus needed in her food, on a dry-matter basis, is:	This percentage of phosphorus will be found on:
3 lb (1·4 kg)	0·55	Superlative pedigree leas
4 lb (1·8 kg)	0·45	The best dairy pastures
5 lb (2·25 kg)	0·35	Good pastures
6 lb (2·7 kg)	0·3	Cut-over woodland, hedgerows, weed patches
7 lb (3·2 kg)	0.25	Good hill pastures
8 lb (3·6 kg)	0·2	Average rough grazing, scrub and hill pasture

It was shown in the American experiments, and must be noted, that though the maximum appetite of the goat is fixed by her physical capacity for food, her actual intake is roughly determined by her current need for nutrients and the form in which she finds them. So the goat giving a gallon (4·5 litres) a day on the best dairy pastures, won't eat much more than 4 lb (1·8 kg) of dry matter a day, though quite capable of eating 7lb (3· 2 kg) a day on good hill pasture.

From the point of view of the goat's health and comfort, it is important that her potential appetite for bulk should be given effect.

The rumen bacteria of the goat, already mentioned in connection with protein requirements, play a very important part in the goat's well-being, and their proper functions can only be performed when the rumen (first stomach) is well supplied with fibrous material. The bacteria break down this fibre into digestible sugars and starches, and release the digestible substances locked in the fibrous cells. In the process they produce most of the heat required to warm the goat and much of her vitamin needs. For a natural inhabitant of mountain tops and windswept cliffs, the goat is singularly ill clad, thin-skinned and lacking in fatty insulation; only when her rumen, which has twice the capacity of that of the sheep, is full of fermenting roughage, can she feel comfortable in cold weather.

Should the goat's ration be lacking in fibre, she will be liable to chill and vitamin deficiency; the rumen bacteria will occupy themselves in breaking down the more expensive and concentrated foodstuffs fed in lieu of roughage, and the cold conditions in the rumen will favour the dominance of disease-producing bacteria such as those associated with entero-toxaemia.

Appetite, however, varies between different breeds of goat. The survival of wild goats in places where the vegetation is too coarse and sparse to maintain any other grazing beast is an indication that a large appetite is characteristic of the species. But goats have long been domesticated, and this character has been modified along with others.

Breeds of goat developed under cold conditions have automatically been developed with large appetites. Low temperatures during early life stimulate more vigorous heart action, which results in the development of wide-sprung ribs. Low temperatures later in life guide the chilly goat to consume the maximum of fibre, for heat. The contrast between the wide-sprung ribs of the Swiss breeds and the flat-sided appearance of the notable dairy breeds from hot countries is self-explanatory.

If it is desirable for economic reasons to feed a concentrated ration—as it may well be in order to cut the labour costs of stall feeding—then a goat of suitable breed type must be chosen. They are not easy to find in Britain, but flat-sided Anglo-Nubians, and goats showing signs of Maltese blood in their appearance, do exist.

PALATABILITY

There is little in the way of green vegetation that the goat will not eat. Its taste is more catholic than that of other grazing stock; its diet includes about 15 per cent more species of plants than the cow's or sheep's. Spines, thorns and bristles are seldom deterrent; stinging-nettles are popular; and even the most tender-mouthed kid will engulf the fully armoured head of a spear thistle with pleasure.

Favourite trees are elm, ash, hazel, willow and oak.

Favourite shrubs are brambles, briars, ivy, gorse and ling heather.

Most favoured among common weeds are hogweed, thistles, seeding stinging-nettles, plantain, cleavers and docks.

The best of the classic crops for goats are maize, mashlum, rape and kale; less usual, but well liked, are sweet blue lupin, artichokes, chicory and comfrey.

Among the roots the potato takes first place, with mangolds and sugar beet second. Swedes and turnips are almost 90 per cent cold water; in winter-time only goats on a very fibrous (and therefore warming) diet will eat them regularly in quantity.

Of hays, goats prefer the leafy and well cured. As they are able to pick out fragments answering to this description from an inferior mass, they do so, and waste in the process most of the inferior mass and a great deal of the best leaf. Clover and lucerne hay is liked by goats and widely used by goatkeepers; but it is expensive to buy, and impossible to grow in many districts; the leaf is excessively fragile and the stem often excessively tough. Hay made from a leafy cocksfoot-and-clover mixture is perhaps the most economical feed; the tough, easily cured cocksfoot leaf encloses the fragile clover in an inextricable tangle, and feeding losses are low. But any really leafy hay will suffice. Hays made from the classic meadow grasses all tend to be minerally deficient for, and relatively unattractive to, the goat. Hay made on a weed patch with the maximum of nettles and docks is always preferable. Dried branches, pure nettle hay, and dried comfrey leaves are luxuries for which only the minority of goatkeepers can find the time, space and sunshine.

In bygone days, gorse was harvested as hay for goats and horses, and ground in whin mills for cattle. The mill, which exercised the bull as well as the horses, consisted of a 6-ft (1·8 m) long truncated cone of granite, which was pivoted in the centre and rolled around a stone circle on which the gorse was laid. The millstones sometimes survive as gateposts. Gorse is good and palatable for goats; the modern hammer mill with a coarse plate overcomes the problem of handling without making the product too dusty.

Of the straws, well-got pea and bean straw are rich in minerals and well liked. Good oat straw is quite as palatable as inferior hay, and for the goat probably has a higher feeding value. Goats eat the head of each oat straw and reject the butt; they eat clean chaff with better appetite and little waste. Other cereal straws are not well liked.

To silage, goats, like cattle, take a ready liking once they have overcome their initial disgust. But goats are more critical than cattle of silage quality, and the losses around the outside of the silage heap are consequently greater when the silage is fed only to goats.

The appetite of the goat for any one concentrate or any one mixture of concentrates, fed in quantities of over 1 to 2 lb (454 to 907 g) per day, soon palls. Goats aren't fussy about the small amount of concentrate they get if fed at lower levels. Strange concentrates in the diet of a goat fed large quantities are always greeted with suspicion. Very high-yielding cows, who have a daily turnover of nutrients and minerals similar in relation to their body reserves, to that of a moderate goat, reveal the same capriciousness towards their concentrate ration. The root of the trouble presumably lies in the fact that no man-made compound or mixture of concentrates is adequately balanced. It is necessary to maintain a variety in the ration, not only to maintain its palatability, but to rectify the errors of one formula by those of the next.

Avoid groundnut cake for goats; it is not unpalatable but has been shown to be poorly utilized. For light rationing, whole oats and bran are liked and suitable. Maize-germ meal and the maize residues from cornflour manufacture are among the cheapest, most efficient and most palatable sources of protein for the goat. Sugar-beet pulp, soaked or dry, is appetizing and rich in calcium; it is an ideal companion to silage, but if fed in quantity requires a phosphorus-rich accompaniment, such as bran. More use might be made of fish meal and herring meal in small proportions; it is mineral-rich and attractive to goats. Heavily fed goats scorning all normal concentrates have been known to consume poultry pellets to their advantage—possibly because of the fish-meal content.

But for the goat, of all animals, to have to have its appetite tempted is a proof of bad management. Bad management is often necessary to win in the show ring with any kind of stock, and the show rings plays its part in serious food production. But the artistry of the goat-feeding bucket is a subject for goat fanciers, and not for goatkeepers or goat farmers, to whom this book is dedicated.

WATER

The goat needs a clean and regular supply. If the available water is hard, allowance should be made for the fact in assessing the animal's need for minerals. 'High-calcium' phosphorus deficiency is a common cause of male infertility in hard-water districts.

The goat derives its heat mainly from the fermentation of roughage in the rumen; in winter-time, cold water is only suitable for a goat with an ample store of roughage inside her.

In the important goat-dairying districts of the south of Spain it is traditional practice to warm the goat's water in the summer sun, and by other means in the winter. The goats are yarded, and fed a concentrated ration. But if the practice has been endorsed by the experience of two or three hundred years of goat farming on the Mediterranean coast, we clearly cannot afford to laugh at the notion of warming the drinking water of any goat fed concentrates in appreciable quantities.

A dry diet with a restricted water supply will tend to fatten any goat. Goats impose such a diet on themselves prior to the breeding season and during early winter, with the effect of reducing their milk yield and increasing their reserves of flesh and fat. An artificially stimulated thirst—brought about by adding salt to the drinking water, or favourite feed—is used to increase milk yield. The same effect is achieved by feeding sloppy concentrate mixtures and dispensing quantities of oatmeal gruel to stall-fed goats. Such techniques strip the goat's body reserves and put them in the milking pail. If the goats are too fat, they may be justified. Otherwise, you have your fun, and pay.

Much controversy surrounds feeding practices, parasite control, etc. The big breeder, striving for perfect conformation, show condition and high yields, is up against digestive and other problems virtually unknown to the owner of a couple of household goats. The first group depends to a great extent on vaccines and drugs, while the small goatkeeper may never encounter anything worse than a minor digestive upset. It is with the household goat that this book is largely concerned and, while the several leaflets and booklets on sale give much useful information, some of the feeding practices will be uneconomic for the many new goatkeepers aiming at a measure of self-sufficiency on a simple basis. The milking goat needs a balanced concentrate ration all the year if she is to fill the pail and maintain condition.

Concentrate mixtures are detailed on p. 162; as an alternative, in view of the high price of flaked maize and linseed cake (both invaluable for condition) they can be modified as follows:

(1) *Economy mixture:*
$1\frac{1}{2}$ parts bruised oats, $\frac{1}{2}$ part flaked maize, 1 part bran, with 1 part locust beans (coarse ground) or 1 part linseed cake.
(2) *Preferred mixture:*
1 part bruised oats, 1 part flaked maize, 1 part linseed cake, 1 part bran (in winter add a handful of grass cubes to one feed).
(3) *Coarse dairy mixture:*
This is now widely used; merchants will make one up to suit your requirements and preferences.

Vary mixtures with bruised barley instead of bruised oats, or split peas instead of beans. A teaspoon of seaweed meal plus mixed minerals, as directed, is generally added to each mixture. Have iodized salt-lick always available. Dairy nuts are not much liked, and can contain undesirable ingredients. The above mixtures are to be fed morning and evening, 2 to $2\frac{1}{2}$ lb (0·9 to 1·1 kg) per head daily. Any change or variation is made gradually: sudden change in diet is a potential hazard.

Ingredients will vary according to availability and costs; but protein, i.e. cake or legumes, must be included. The most important single factor in the diet is well-got hay, preferably a sample with some clover and herbs in variety; it must be sweet-smelling and leafy, or the goats will refuse it. Vary in winter with first-class oat straw, or feed a sheaf of oats once or twice daily. Clean-got oat straw is very popular, makes a change and saves hay.

Bed the goats on barley straw for preference; fresh oat straw stains. A good bed of straw topped daily results in splendid compost, as well as keeping the goats warm in winter. As a useful winter midday feed for milkers, baked or boiled potatoes, broken open and given a sprinkle of grass meal or bran make an economic feed, safer and more digestible than soaked beet pulp. Water, preferably warm with salt added, is necessary; kids should be offered it at an early stage. The big milker is always the big drinker. The concentrate diet is supplemented by practically everything from the garden not eaten by the family.

Free-range grazing and browsing is the ideal system provided a frequent change to clean ground is possible. If there is no opportunity for free-range grazing, goats can be successfully yarded and hand-fed (see the section on zero grazing, p. 315). Under range conditions, worming can be reduced to Thibenzole spring and autumn, vigorous kids having their first treatment at about a year old. Garlic is widely used for all ages as a blood purifier and preventive against parasites and disease. Given in tablet form, there is no taint to milk; in natural form, feed directly after milking. Two or three tablets twice weekly as routine practice is recommended. If goats must be turned out constantly on the same small area, inevitably leading to a build-up of parasites, more stringent routine worming and other measures, as advised by the vet, will be necessary to maintain health—in fact literally to keep them alive.

On free range with good browsing and grazing the kid needs no concentrates, indeed is much better without; the widely held belief it will never eat concentrates unless given them early in life is absurd. During winter, especially at the turn of the year, a healthy kid will greedily eat its balanced ration of 1 lb (454 g) daily, and ask for more. As a goatling on spring grazing and browsing, she will again be better with no concentrates, able to keep on steadily growing till July, when supplementary feeding is indicated to compensate for the drop in nutritional value of grass, etc., and to have her in prime condition for the breeding season. An enormous amount of money is wasted, and robust constitutions risked, by feeding concentrates to kids and goatlings, which could be avoided with ease—provided, of course, they have the advantage of free-range management, or get plenty of good fresh pickings from the garden or from green crops grown specially for goats. Should a big early-born kid with good bone be destined for mating at eleven months, she will naturally need extra feeding; but in the ordinary way she is likely to do better in later life, and maintain a higher standard of health, if she develops gradually.

Earlier editions of this book have been widely criticized for advocating the small amount of 2 to $2\frac{1}{2}$ lb (0·9 to 1·1 kg) concentrates. Most goatkeepers, and certainly all those feeding for high milk yields and peak show condition, feed considerably more. Nevertheless, practice has consistently shown that on this low level, yields remain high and health excellent. This moderate feeding of concentrates needs to be linked with wide free-range grazing in summer and, for example, kale, roots and well-got home-grown tripoded, herbal lea hay and oat straw in winter. It could well be a starvation diet, as the critics suggest, if not supplemented by generous feeding, either range or yard, of cut green stuff, branches, roots and any other supplement.

It is usual at shows to see the goats diving into enormous quantities—as much as three-quarter pails full—of concentrates, fortified with high-protein cake; as well as huge quantities of soaked beet pulp together with its liquid. That this is not a diet that ensures health is quite evident for their owners must have the hypodermic always at the ready, at home and abroad. High milk yields, large size and never a rib showing are the criteria with big breeders—a game of skill which does not appeal to all of us, and expensive in more ways than one.

Chapter Seven

Feeding Practice

Under farm conditions, the rationing of the goats will be done almost entirely by the goats themselves. It differs in no way from the general procedure of rationing dairy cows, except that the pastures will usually cost less or nothing to grow, and the amount of bag feeding required will be less per gallon of milk and more easily measured.

We have now reached the end of a small instruction booklet on operating a highly efficient small milk-producing machine known as 'The Goat'. We have dealt with the subject of maintenance and running repairs, with fuel consumption and the suitable types of fuel; we have issued the necessary reminder to keep it topped up with water. All that 'The Goat' salesman has to do now is to point out to the customer that he will find at the end of the book all the requisite information about types of fuel available in the district (Appendix 2). The customer must, of course, decide for himself how much milk he wants from the machine and fuel it accordingly. With a complacent farewell, the salesman may now get into his delivery van and sail away.

This is, in effect, just what the conventional exponents of 'scientific rationing for milk production' do. If they don't exactly sell their customers a pup, they certainly sell an animal of some kind under the false pretence that it is a machine.

The swindle is not immediately apparent. If you buy a goat which gives a gallon (4·5 litres) of milk a day, and which the railway consignment note shows to weigh 140 lb (63·5 kg) and you feed her in accordance with the formulae given in this chapter . . .

i.e. for maintenance $0.9 \times 1.4 = 1.26$ lb (571 g) starch equivalent
and $0.09 \times 1.4 = 0.12$ lb (54 g) digestible protein
for production (1 gallon) $= 3.25$ lb (1.5 kg) starch equivalent
and 0.5 lb (227 g) digestible protein
All incorporated in $6 \times 1.4 = 8.4$ lb (3.8 kg) dry matter and in
palatable form

. . . then you are not likely to experience any serious trouble for some time. You will, of course, cut down her ration as her yield declines in the later stages of lactation. In early autumn she will show some signs of being unsatisfied with her diet, and will raid the food bin if she gets the chance. By Christmas she will have little appetite even for a small ration, and will probably go dry. During late January and February she will show further signs of being unsatisfied with her ration; when she kids in March she will yield only 6 pints (3.4 litres) a day instead of 8 (4.5 litres). You will then become very angry with 'scientific rationing', decide that you have been underfeeding the goat (which is quite true) and proceed to give a much more concentrated ration. The effect on the goat's yield will be slight and transient; the effect on her health may be greater. Autumn will probably find you trying to bolster up a declining yield with ever-mounting heaps of concentrates. Usually during the first half of September, but often nearer Christmas, you will end your experiment in 'scientific rationing' as you fill in the grave.

The Metabolic cycle

There are empirical 'wrinkles' which avert some of the consequences of mechanical rationing; most helpful is the instruction to 'steam up' a goat before kidding; that is, to provide a full production ration during the last eight weeks of pregnancy. Even this job can be done more efficiently if you understand the reason for doing it, and if the process is part of an all-year rational scheme of feeding, which treats the goat as an animal, and not a machine.

The crowning fallacy of mechanical rationing is the notion that the goat's expenditure of nutrients should exactly balance her income on a day-to-day basis. In reality, like so many flourishing agricultural enterprises, the goat lives on her overdrafts for a large proportion of the year, and it is only during the last few weeks of pregnancy that the balance is struck and her solvency tested.

The advantages of the arrangement are easily seen. During the late summer, autumn and early winter, the goat's milk production is relatively light and her digestive capacity is not fully employed in supplying her udder. Given the opportunity, she can at these periods lay up a reserve in her body, on which she can draw during the following spring and summer when the demands of her udder will usually exceed the throughput of her digestive system. Like a good motorcyclist, she takes her corners before she comes to them, so that at the point of greatest danger there is less stress.

The agency by which this arrangement is regulated is the activity of the thyroid gland, which controls the goat's metabolic rate—the speed at which her body processes work, her activity and appetite. The exact timing of the metabolic cycle does vary with the timing of the breeding season, which begins nearer to midsummer as the goat lives nearer to the Pole (North Pole in the Northern Hemisphere and South Pole in the Southern Hemisphere). This shift in timing makes it exceedingly difficult for goat breeders in Scotland to record milk yields as high as those obtainable farther south, because the peak of the metabolic rate of goats in Scotland is reached at a time of year when natural feeding is poor.

Metabolic rate reaches its lowest level in early autumn, about three weeks before the onset of the breeding season. The most obvious indication of this stage in the cycle is the sudden drop in yield of the milking goat, which continues for about a week and then stops, the yield remaining at the lower level as regards quantity, but improving in quality. No purpose is served in trying to oppose this fall in yield. Immediately the fall becomes apparent, introduce, or allow the goat access to rougher and drier forage; the drier ration will assist the goat to divert its food into building reserves for next lactation.

As soon as the yield settles down at the new level, the appetite of the goat increases rapidly. Feeding during the subsequent month is specially important for the goat's fertility, should enable her to make substantial weight gains and fit her to face the chills of autumn. The diet should remain dry and rich in fibre; the energy value of the ration should rise to that of a full production ration at the height of the breeding season, but the protein content of the ration can be relatively low—1 part digestible protein to 8 parts starch equivalent for a goat whose yield at the time has fallen to under 4 pints (2·3 litres) or a 1:7 ration for heavier milkers.

After the peak of the breeding season is passed, the yield and appetite of the goat declines until midwinter. Feed in sympathy with this decline, maintaining a dry type of diet relatively rich in energy (carbohydrates) and fibre.

The recurrent periods of 'heat' during the breeding season affect different goats in different ways. In some goats, the secretion of hormones is so powerful in relation to the activity of the thyroid that milk yield almost ceases; but in a goat with a very active thyroid, the opposite effect may be produced. In either case, it is an inherent characteristic of the goat which diet does little to alter.

Whether the goat milks through the winter or dries off before midwinter depends on the activity of the thyroid, and on whether she is in kid. While the activity of the thyroid is largely an inherent characteristic, maximum activity (and therefore productivity) for any one goat is secured by feeding in full sympathy with the fluctuations in metabolic rate. The greater liberty the goat has to select her own ration in type and quantity, the more exaxt will be the sympathy. Man-designed rations are at best a necessary evil.

The common, and fatal, error is to adopt a feeding policy designed to secure a high uniform level of production throughout the year. To do so is to fight a pitched battle with the glandular rhythm of the goat and the whole patterning of its vital processes; to turn the goat into a battlefield for an idiotic contest between the veterinary profession and the agents of disease.

Winter Feeding

From midwinter until early spring, the metabolic rate of the goat rises steadily to its maximum level. The goat feeder's job is to organize a crescendo of quantity, quality and palatability in the feeding racks to support the natural upsurge of productivity and add impetus. But some caution is necessary if the goat is in kid, and especially if she is naturally capable of high yields.

Winter production is a drag on the development of condition and vitality if the goat is also supplying the needs of growing kids in her womb. In any goat that milks past midwinter, the yield will increase as the days lengthen; if the early increase is given much encouragement, a good milker will milk till she kids again; the strain on her resources is excessive and her performance in the subsequent lactation will suffer.

To dry off a goat of a high-yielding strain is not easy; mere starvation gives unsatisfactory results. But there is no objection to restricting water intake and making her diet as dry as possible over a short period. A teaspoon of cod-liver oil per day will give assistance. By these means, and by milking irregularly and without stripping, every effort should be made to ensure that the in-kid goat has ended her last lactation at least eight weeks before she is due to start her next.

The goat who is in kid but not in milk after midwinter also requires cautious feeding during the early part of that year. A high level of feeding throughout pregnancy makes kids that are too big to be born with ease: if the diet is also high in minerals, the bones of the kids will be too hard and solid at birth. A gently rising, dry, and fibrous diet with a rather low protein content is indicated. The kids grow very slowly during the first three months of pregnancy, and their food requirements are insignificant.

'Steaming-up'

During the last eight weeks of pregnancy, the growth of the kids is rapid, and the metabolic rate of the goat rises rapidly. Through this period, the content of the ration should rise to the level of the full production ration—that is, the theoretical requirement of the goat when giving the peak yield that is expected of her—and the level should be reached a week before the goat kids. At the same time the dry nature of the ration should be changed, and the sappiest materials available supplied in lieu.

It is hard to design a sappy, rich diet in early spring, when the marrow-stem kale is reduced to blasted stumps and most of the roots have been used to make the ration too succulent in early winter. But with planning it is possible: 'Hungry Gap' kale will be fruitful at this time, and so will Russian kale. Mangolds, fodder beet and carrots can be stored for this occasion. In southern districts, comfrey will be making succulent growth. At the worst, sugar-beet pulp can be fed soaked instead of dry. But in the effort to increase water intake, the objection to cold water must not be forgotten.

For three or four days after kidding, the level of the diet should be lowered and made more fibrous, to minimize the shock to the goat's udder and metabolism of the sudden flow of milk. Thereafter it should return to the previous top level and stay there for three or four weeks. After the lactation is well established, the goat can safely be fed a little below her theoretical requirements. She should be carrying the makings of many gallons of milk on her back and chest, and should be allowed gradually to dispose of the burden into the milk pail. During the first four months of lactation the milking goat should lose weight steadily.

This system of feeding is less revolutionary than most goatkeepers will believe at first sight. In essence it is the system developed and used in the dairy cattle herd belonging to Boots Ltd, under the supervision of Mr Stephen Williams, M.Sc., and which has been successful for many years. This herd is characterized by regular calving, complete freedom from chronic contagious mastitis, a marked freedom from digestive disorders and complete freedom from tuberculosis. It can claim a 305-day herd average of 940–1,000 gallons (4,273–4,546 litres), a pattern of progressive lactations—850, 950, 1,000, 1,100, 1,200 gallons (3,864, 4,318, 4,546, 5,001, 5,456 litres)—and a number of 10,000-gallon (45,460 litre) cows.

   The system suggested accords well with that recommended by the late Mr Newman Turner for cattle—and it is surely remarkably solid ground upon which Boots Ltd and Mr Newman Turner can find a common footing.

   The natural metabolic cycle of dairy cattle has been messed up by winter breeding, and their daily dry-matter capacity extends only from $2\frac{1}{2}$ to $3\frac{1}{2}$ per cent of bodyweight. For goats, who have a restricted breeding season and a far wider range of capacity, the system clearly has greater merits.

   The main merit of the system lies in the fact that it spreads the work which has to be done by the goat's digestion more evenly over the year. This makes the daily task of digestion sufficiently small, to enable the goat to obtain by far the greater part of her food requirements in the form of fresh greens and roughages which she can ration for herself.

Frequency of Feeding

When a goat cannot obtain what she wants, when she wants, not only have we the risk and labour of rationing her, but we must see that her ration is split into small and frequent feeds. Feeding with the flock on free range all day, the goat eats a little, cuds a little and eats a little again; shut in a yard or stall with other goats or alone, or allowed out to graze for a restricted period, the goat stuffs all she can while it is there. The difference in the food value of a given weight of fodder has been shown to be twice as great if the day's ration is divided into five parts offered at intervals during the day than if it is all fed at one time or, to put the matter in a more frightening way, the goat needs twice as much food per gallon of milk produced if the food is fed at a single feed than if it is fed at five separate feeds—what is more, she will eat it. This is fairly well understood in normal stall-feeding practice; but it tends to be forgotten when goats are put out to graze for restricted periods.

The reason for the importance of the 'little and often' principle is that the rumen bacteria attack the contents of the rumen as long as it is there. At first, their attack results mainly in the conversion of fibre into digestible sugars and the release of the digestible substances locked in the fibrous cells. But if the food stays long in the rumen, the digestible sugars and proteins so released are broken down into heat and gas. With frequent feeding, small quantities of food pass rapidly through the rumen and the nutrients in them are digested by the goat; with massive single feeds the food remains packed in the rumen for a long period and sustains heavy losses.

Goats put out for two or three hours' grazing and brought in again benefit little more than if they were out for a quarter of an hour.

Feeding for Butter Fat

The capacity to produce milk with a high percentage of butter fat is inherited, butter-fat percentage being one of the easiest breed characteristics to fix. But the ability of an individual to exercise her inherent capacity for butter-fat production depends on several other factors. To understand the situation, it is necessary to take a closer look at the goat's internal economy.

The goat's rumen (first stomach) is a fermentation vat, where all she eats is fermented by bacteria and changed into the substances which the goat actually digests and absorbs. Though some food may pass through too quickly to be changed—dairy nuts on an empty stomach, for instance—most food is utterly changed before it is presented for true digestion, or for absorption through the walls of the rumen itself. Energy foods are absorbed in the form of fatty acids; the sugars and starches, which constitute most of the energy food in green fodder and concentrates, are changed into butyric and propionic acid; the cellulose, which comprises most of the fibrous matter in the diet, is changed into acetic acid. The goat depends upon the supply of acetic acid to make her butter fat.

To realize her inherent potential for butter-fat production, the goat needs, not only an adequate proportion of fibrous fodder in her diet during lactation, but also to be so reared and managed as to develop to the full the capacity of her digestive organs, so that she can accommodate so bulky a diet in sufficient quantity.

The average butter-fat percentage recorded for a milking goat over a complete lactation will reflect her inherited potential, modified by the way in which she was reared, the degree of maturity at which she kidded, and the nature of her diet during lactation.

Rearing methods affect butter-fat percentage by modifying the development of the digestive organs, especially the rumen. The ultimate size of the rumen is stunted by feeding more than 4 pints (2·3 litres) of milk a day to a female kid, or feeding substantial quantities of milk to kids over four months old, or feeding more than 1 lb (454 g) of concentrates a day to kids or maiden goatlings. The stunted organ cannot accommodate enough fibrous fodder to do justice to the goat's inherent capacity for fat production.

The degree of maturity at which a goat kids limits the size of rumen, her capacity for fibre and her butter-fat potential, during that lactation. There is little growth in a goat's frame and digestive organs during the last two months of pregnancy and the first three months of lactation; but if adequately fed during the later stages of lactation, a young goat continues to develop for several years. Goats who start breeding as kids may, indeed, be permanently stunted, and their butter-fat production permanently impaired. But the practice of 'running through' (extending the lactation over two years, without breeding) helps productive young goats to complete their growth and develop their butter-fat potential to its full. A goat's butter-fat production should continue to improve until it is five years old,

The effect of diet on butter fat is not confined to the need for an adequate proportion of fibrous food. As explained in the previous chapter, lack of salt or cobalt in the diet results in diminished appetite, and impedes the fermentation of fibre in the rumen, so restricting butter-fat production. Worm infestation raises the salt and cobalt requirement of the goat and accentuates the effect of dietary deficiency. Vitamin B_{12}, synthesized from cobalt by the rumen bacteria, is used by the goat in the metabolism of butter fat. A slight deficiency of this vitamin may result in lowered butter-fat production, imperfect metabolism of acetic acid to butterfat and bitter-tasting milk. Such a deficiency may be caused by worm infestation, lack of fibre in the diet, lack of cobalt, or all three misfortunes concurrently.

Of course, the more milk a goat gives, the more fibre she needs in the diet to sustain a maximum percentage of butter fat. When the proportion of fibre in a diet rises above a certain level, milk yield falls. For any individual goat, there is a point at which milk yield can only be increased at the expense of butter fat, and butter-fat percentage only improved at the expense of milk yield. The level at which this occurs depends mainly on the size of the goat's rumen.

The embarrassing fall in butter-fat percentage which occurs in the show yard, and the chronically low butter fats when goats go out on to lush spring grass, can both be prevented by a substantial feed of hay, morning and evening; this seldom has any adverse effect on yield.*

In feeding for butter fat, it is well to remember that there are some variations in butter-fat percentage that have nothing to do with diet. There is the seasonal variation, illustrated in Fig. 12, which is determined by the basic metabolic cycle of the goat, and is practically unchangeable. Variations in milking routine may be reflected in butter-fat percentage. Equal intervals between milkings tend to maintain a steady proportion of butter fat in the milk. If the intervals are uneven, in the milk taken after the long interval, the percentage of butter fat is reduced; in the milk taken after the shorter interval it is correspondingly increased. The first milk drawn at a milking is low in butter fat; the last milk—the 'strippings'—rich in butter fat. Occasional failure to 'strip out', and extract the last available drop, lowers the percentage of butter fat at the milking at which the failure occurs; but the loss is recovered at the next milking; regular failure to 'strip out' does not affect the total yield of butter fat.

---

* Low butter fat at shows may be due, also, to anxiety inhibiting 'let-down' (p. 260).

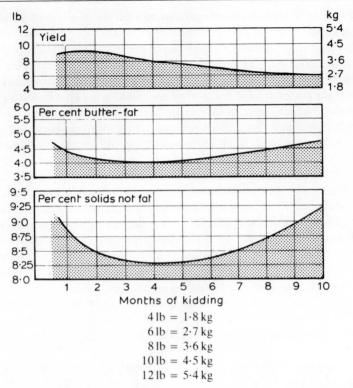

lb
kg

12 — 5·4
10 — 4·5
8 — 3·6
6 — 2·7
4 — 1·8

Yield

6·0
5·5
5·0
4·5
4·0
3·5

Per cent butter-fat

9·5
9·25
9·0
8·75
8·5
8·25
8·0

Per cent solids not fat

1 2 3 4 5 6 7 8 9 10

Months of kidding

4 lb = 1·8 kg
6 lb = 2·7 kg
8 lb = 3·6 kg
10 lb = 4·5 kg
12 lb = 5·4 kg

Fig. 12. Lactation curves

The Collapse of Productivity

The average yield of goats at agricultural shows dropped by $12\frac{1}{2}$ per cent in the four years which followed the effective derationing of animal feeding stuffs in 1949. So sudden a fall excludes the possibility that bad breeding policies were mainly to blame, and its timing points to the major cause.

During the years of the Second World War and 'austerity', the official concentrate ration for a goat producing a reasonable yield of marketable milk was seldom inadequate for a goat which had access to plenty of bulk food—grazing, roots and hay. The allowance for kids and goatlings, and for milk used for kid feeding, was very much more restrictive; it was the feeding of young stock which changed most drastically with the availability of concentrates.

The hindsight offered by Fig. 13 allows us to recognize that the restriction of milk and concentrates for kid feeding resulted in higher yields, with a higher percentage of butter fat, in the mature goat. As we have shown above, full development of the rumen, necessary for satisfactory butter-fat production, is impeded by feeding excessive quantities of milk and concentrate to kid and goatling. Although *percentage* of butter fat and yield of milk often vary inversely to each other, *total butter-fat production* and yield of milk usually rise and fall together. To put the matter in another way, both butter fat and protein are essential constituents of the species and of the individual goat; within these characteristic proportions, the percentage of protein or butter fat in the milk varies a little with the composition of the diet. But if the diet fails to provide the makings of enough protein, or enough butter fat, to maintain the characteristic pattern, then the yield falls.

British Goat Society show regulations heavily penalize a low

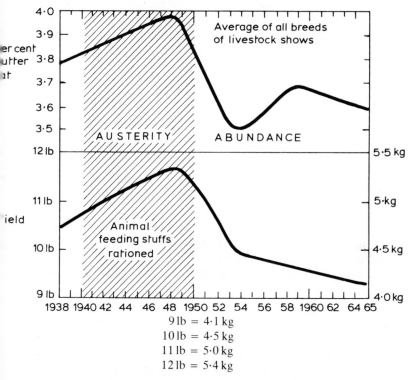

Fig. 13. The effect of concentrate feeding, 1938–65

butter-fat percentage, and encourage exhibitors to breed for higher butter fat. But unless better kid-rearing methods permit better rumen development and better butter-fat production, the higher the butter-fat characteristic of the goat, the smaller her yield will be, as Fig. 13 so tragically records. British Goat Society show regulations also endorse the unparalleled practice of holding kid classes, where the most overfed kid, with the most stunted rumen development, usually wins. The aesthetic effect of kid classes is delightful; their effect on productivity is utterly malignant.

It may seem unwise to base so much argument on show-yard records, and to disregard the evidence of the official milk records. In fact the records of the few goat flocks which are officially recorded confirm the general trend shown by the show records. The show records are preferred because they cover a wider spectrum of goat breeders, and because show yields are produced in public by public methods. In some recorded herds, all the milk which cannot be bottled into kids is fed back to the goat who gave it, and the official records refer to milk circulation rather than milk production!

In analysing the 3,000, or so, show records which went to the making of Fig. 13, it was striking to note how many of the high-yielding goats of the 1950s were unpedigreed goats, or goats with an unregistered dam. These were the last of the cottagers' goats; reared cheaply on bulk foods, they had a good capacity for bulk; by the time they reached the show ring, their yields, though high, were low in butter fat, signifying the over-concentrated diet favoured by the show-ring enthusiasts. By 1965, this type of goat was not to be found. Presumably, the show ring has debauched the last of them.

Breeding stock for goat-improvement schemes all over the world come from the mountains, where low temperatures and tough, frost-resistant herbage stimulate the development of bulk capacity from birth to maturity. If we wish to restore British goat productivity to its former glory, we must create mountain conditions in the goat house, as we were reluctantly forced to do so by wartime and post-war shortages.

THE FEEDING OF YOUNG STOCK

The milking capacity of a goat is in the main an inherited characteristic, dependent principally on the inherent activity of her thyroid gland. In feeding for production, we feed in sympathy with the fluctuations of the goat's natural metabolic cycle, to secure maximum thyroid activity and maximum use of the goat's milking capacity.

But in rearing young stock, the best that good rearing can do is to condition and modify the development of the kid's digestive organs to suit it for the diet it will receive in later life. The natural growth rate of kids fluctuates in sympathy with the metabolic cycle of adult goats. Though we feed the kid in sympathy with these natural fluctuations in growth rate, there is no good purpose served in attempting to achieve maximum thyroid activity or the rapid, sappy growth that such activity produces in the young goat. Rapid growth always endangers the potential productivity of the kid in later life.

The principal feeding problem of milking goats concerns the maintenance of a correct mineral balance. This is true for even the more moderate milkers; high-yielding goats need a highly mineralized diet on which they have to maintain an extremely precarious balance between calcium deficiency, phosphorus deficiency and iodine deficiency. The best insurance they can have against disaster in the process is the possession of substantial mineral reserves banked in their bones, on which they can draw in emergency. A well-mineralized skeleton is made during the first eighteen months of life or never at all. The potential value of a kid of generally good type lies largely in the invisible minerals secreted in its bones.

The job of the kid rearer is to build well-mineralized skeletons; the job of thyroxine in the blood of the goat is to divert the dietary supply of minerals away from the skeleton and into the soft tissues. So we must curb excessive thyroid activity—that is, maintain a normal growth rate. The more milky the pedigree of the kid, the more urgent and difficult this task becomes.

The contrast between this advice and the practice of many leading pedigree breeders is so violent that it is necessary to emphasize that it is based on massive and unanimous scientific support from the 1930s onwards. (Owen, *British Journal of Nutrition*, 6, 415; Fairbanks and Mitchell, *Journal of Nutrition* (1936), 11, 551; Zucker and Zucker, *American Journal of Physiology*, 146, 593; Duckworth and Godden, *Journal of Physiology*, 99, 1; Duckworth, Godden and Warnock, *Biochemical Journal*, 34, 97.)

The normal growth rate of kids represents a daily weight gain of about $\frac{1}{2}$ lb (227 g) per day for the first six months of life. Such normal growth can be obtained on the following régime:

Wherever possible, the kid should be allowed to suck its mother for the first four days of life. The arrangement is as helpful in conditioning the mother's udder as it is in giving the kid the best possible start.

If the kid must be bottle-fed at birth, it must be fed with the milk of its mother or that of a freshly kidded goat. It matters relatively little, from this point of view, if the mother has been pre-milked, or only finished its previous lactation a day or two before kidding. The importance of the kid receiving milk from a freshly kidded goat lies in the rich vitamin-A content of such milk, and in its wealth of antibodies to give protection from disease. Such milk is also laxative. In spite of pre-milking, the milk is enriched with vitamin A only by the act of kidding. The milk of the goat will permanently contain antibodies to all or most of the infections to which she is exposed; the supply will be less rich in the milk of a pre-milked goat, but still adequate; but the laxative properties of the milk are reduced by pre-milking, and the kid fed on it may be given a teaspoonful of medicinal paraffin one hour after each feed for the first twenty-four to thirty-six hours. If given with the feed, the liquid paraffin will reduce the vitamin-A potency of the milk.

For bottle feeding a kid during the first few days of life, a polythene baby-feeding bottle with baby or soft lamb teat is most convenient. The flexibility of the polythene bottle enables the feeder to 'let down' the milk, so that the kid is not embarrassed with the effects of the vacuum its sucking creates.* The kid should be allowed to take as much milk as it wants at six feeds in the day. The bottle should be sterilized after each feed, and the goat milked anew for each feed.

At four days old, most kids will be weaned from their mothers on to the bottle. Purely from the viewpoint of rearing the kid well, there is no objection to the average kid being allowed to suck its mother all the summer; but exceptionally well-bred kids are better bottle-fed, or they may grow excessively fast. Though rapid growth on a milk diet in which the minerals are well balanced is unlikely to

* More recently there has appeared on the market the 'freflo' nylon baby bottle, which has every virtue the kid feeder could wish for—a teat which regulates the flow, prevents vacuum, and cannot be swallowed by a kid, a wide neck for easy cleaning, and made of unbreakable material.

produce mineral deficiency in the skeleton, it can only be achieved by consuming such large quantities of milk that the digestive system of the kid becomes distorted. The second stomach of a calf has been shown to increase in size by fifty-four times in the first three months, when the calf is fed a normal diet including roughage. On an all-milk diet it increases in size only three times in the same period. A kid fed abnormally large quantities of milk, direct from the mother or out of a bottle, will not have a normally developed rumen. The longer the milk diet is maintained, the more permanent the deformity will become. In effect, protracted feeding of a big milk ration produces the handsome-looking goat who is unable to realize her innate capacity for milk production.

Four pints (2·3 litres) of milk a day, in four feeds, is adequate for any female kid. A billy kid may have an extra pint (0·6 litres). This level of feeding will usually be reached when the kid is a fortnight old, if the feeder allows the appetite of the kid to guide him. A wine bottle or calf feeder replaces the baby bottle at this state.

At the age of a fortnight kids will want to nibble soil to satisfy their needs of iron and copper, in which the milk is deficient. There is a certain risk that, if allowed to nibble the soil round about the goat sheds the kid will pick up an infestation of the tapeworm, *Monezia expansa*, which will rob it of calcium for the next six months. To minimize this risk, the kid should have access to a mineral lick or a box of clean earth; we will naturally want to give it the best earth we have got, but the best for the purpose is second-spit subsoil or the molehill the kid will usually select for itself—and certainly not compost.

At this stage, too, the kid will start nibbling hay, and should be allowed to do so. But hay nets have hanged a lot of kids. The hay is best placed on the ground and the wastage accepted.

The introduction of milk substitute can be started at this age if the goats' milk is urgently needed elsewhere. Perfectly satisfactory results seem to be achieved by the use of these substitutes, if care is taken to follow the makers' instructions exactly and to carry out the introduction little by little. No rearing method can be judged by appearances; when a number of goats able to sustain yield of over 300 gallons (1,364 litres) a year, without mineral deficiency or digestive troubles, have been reared on milk substitutes, it will be time to endorse the claims of the makers. One thing is sure: these substitutes contain none of the antibodies to the infections prevalent in the district which are conveyed to the kid by natural milk. Rigid adherence to the routine of sterilizing the feeding bottle after each feed is therefore necessary. If the kid scours, miss a feed and reduce the proportion of substitute.

On milk feeding, scours result mainly from the use of dirty bottles or the offering of cold milk. Milk should always be given at blood heat. Where facilities for warming milk immediately before the feed are lacking, fill a large Thermos with boiling water and heat the bottle in it when the time comes. On no account should the milk be kept warm over a long period.

From two weeks until four months old, the kid needs no more than her 4 pints (2·3 litres) of milk a day (which provide the food equivalent of ¾ lb [340 g] of concentrates), and access to sunshine, fresh green food, salt and water. Skim milk may be used to replace part of the milk ration at three months, in which case a handful of oats or flake maize may be offered in lieu of the fat.

At four months, milk may be reduced from 4 pints to 3 (1·7 litres) in the day; at five months, from 3 to 2 (1·1 litres); and at six months, from 2 to 1 (0·6 litres). After a further fortnight, bottle feeding should be stopped. As the milk is reduced, it should be replaced by good-quality roughages if available, and preferably not by concentrates.

Though the growth rate of the kid may slow down as a consequence the digestive organs will develop their capacity for bulk and the appropriate bacterial population to deal with it.

Goatkeepers tend to spend more time than necessary on kid rearing—partly, perhaps, for the sound reason that it is a delightful occupation, but partly too because of a misguided determination to see that the future generation get the best of everything. Owing to the concentrated and highly digestible nature of their basic diet, milk-fed kids can, in fact, thrive on rougher fodder than adult stock. They will eat that portion of the hay which their mothers reject, not out of youthful indiscretion, but because their needs are more modest; in the same way the kid will be chewing twigs while her mother is eating leaves from the same shrub. The only fodder which is suitable for adults but unsuitable for kids is that which is too nutritious for the latter.

Most of the traditional anxiety to separate kids and adults at pasture, and to provide the kids with special fodder, is the fruit of the bitter experience of running young kids and adults together on good grass. Young pasture grass is too rich in nutrient and poor in minerals to be a safe staple diet for adult goats; for kids it is definitely dangerous in quantity. When wet, it forms a cold, tight-packed indigestible wad in the kid rumen, as many a sad post-mortem examination has revealed.

We shall greatly improve our pastures if we include in them a substantial proportion of deep-rooting 'herbs' such as burnet, chicory, dandelion and yarrow; these contain more minerals than grass does and make the pasture healthier, but no less productive, for milking stock—goat, cow or ewe. On such a pasture kids may be safely put out to graze at two or three weeks of age if the 'herbal' content is making a good showing—just as they can go out on scrub browsing. But if the pasture is of the conventional grass-and-clover type, kids up to at least three months of age must receive a good feed of hay before they are put out, and must never be put out if the grass is wet.

The heaviest mortality in kids occurs during their first autumn and in the following spring. The fatal disease is usually either pneumonia or enterotoxaemia, in many cases aggravated by worm infestation. All these agents of disease are normally present in the body of the healthy kid.

Disease is no more caused by specific bacteria or parasites than Christmas is caused by the overloaded postman. You can suppress the symptoms of disease or Christmas by obstructing the bacteria, parasites or postmen, but in so doing you neither remove the cause nor prevent the recurrence of symptoms. The cause of disease is the condition which allows the agents of disease to become over-active.

Kids and lambs develop heavy worm infestation in autumn if they lose much weight after weaning; that is, if their digestive system has become too dependent on milk, or if the after-weaning diet is unsuitable. They become heavily infested in spring if available feeding fails to improve in quality in sympathy with their increasing metabolic rate and appetite; that is, if they are either half-starved all the time, or fed so generously through the winter that it is impossible to pack more food into them in the spring. Pneumonia affects mostly the leaner kids in these categories; enterotoxaemia tends to affect the fat and thriving ones—especially the unweaned kid, fed enough concentrates to carry the rapid growth of summer on into the winter. Cases of enterotoxaemia and pneumonia are often precipitated by sudden changes in diet or weather; but the same changes do not affect all kids alike. They affect the kids who had too much milk in their first summer, or too much concentrate in their first winter, even more than they affect the starveling and weakling.

The ability of the kid to thrive on bulk foods and roughages alone is not fully developed until it is about nine months old. The March-born kid, weaned at about four months old, will thrive on the common run of grazing and browsing into the autumn, for the gradual hardening of the fodder matches the kid's developing powers of coping with it. Later-born kids, weaned at the same age, need to have the pick of the bulk foods available if they are to maintain progress. Cabbage, kale, good stubble and aftermath grazing are adequate feeding for the late kids after weaning. If they must put up with lower-quality bulk foods they will require a supplement of concentrates, preferably those, such as oats or bran, which have a moderate fibre content. Good hay, with a maximum of 1 lb (454 g) per day of concentrates (7 parts starch equivalent to 1 part digestible protein), will see any kid through the winter until the beginning of February; the more foraging and exercise it can receive in the meantime, the better. From February on, feed a wetter diet, using 'hardy green' turnip tops, fodder beet, sugar-beet pulp, 'Hungry Gap' and 'Asparagus' kale, etc. At the same time step up the protein in the ration to a 1:6 ratio, until the new grass comes; then drop the protein content, and phase out concentrates altogether; but go on feeding hay, morning and evening, until the first flush of grass is over.

Once safely into its second summer, the kid ceases to be a problem child and becomes a goatling, needing nothing but her fill of fresh fodder, salt and sunshine, until she joins the ranks of the breeding adults in early autumn. Overfed goatlings are liable to summer mastitis; but there is no call to feed concentrates to a goatling until she is in kid, and no possibility of a goatling being overfed on bulk foods.

Sometimes a kid is purposely or accidentally served during her first autumn or early winter. This is most likely to occur to a growthy kid, highly stimulated by its diet. In many ways it is the best thing to happen to such a kid; as goatlings they tend to be over-fat and minerally imbalanced, and difficult to get in kid. But all kids are sexually potent at a young age, and a normally developed kid can easily be served at six months or younger. In any case, the kid who has to build its own bone and body and that of the foetus, and at the same time to lay up reserves for lactation, is going to be hard-pressed for calcium, phosphorus, vitamin D and general nutrients. But the same pattern or feeding must be followed, whether the kid is served or not. Only from the beginning of February is it safe to feed the immature first kidder on a diet rising to substantially higher levels. During the earlier part of her pregnancy the most that can be done is to ensure a highly mineralized diet with a vitamin-D supplement. The main danger in feeding the immature breeder lies in over-generosity; it is sufficient that her ration should rise during the last week of pregnancy to a maintenance-and-production ration for her actual weight and 6 pints (3·4 litres) of milk. After kidding, this ration should not be increased, except in rare cases; the lactation

Plate 9. Toggenburg kid, Murrayston Bonimaiden.

should be curtailed as soon as possible if good subsequent performance is to be expected.

Billy kids should always be serving when they are about six months old; if they will not serve then, they never will and should be destroyed without more ado. Consequently, and to meet their rather faster growth rate, billy kids get more milk (5 pints [2·8 litres] a day) in their bottles; may be fed the full amount up to the age of six months; and continue to receive 2 pints (1·1 litres) a day for the duration of the breeding season. Otherwise male-kid feeding follows the lines already indicated for their sisters.

It is common practice to restrict the activities of valuable pedigree billy kids to about six services in their first season. The practice does no harm, as the kid, if not adequately employed, will masturbate continuously. Though the effect of a lot of work is to reduce the male kid's condition during the autumn and early winter, it has not been shown to reduce his subsequent fertility; unless allowed to accumulate a heavy worm infestation while in low condition, a rising level of feeding in early spring will restore the kid to condition. After summering on an ad lib supply of fresh fodder, such a male is more likely to make a vigorous and fertile sire than one which has never lost its puppy-fat. The idea is to give the billy kid enough work and food to keep him in good lean condition.

Male goats are always available in ample supply, and are very cheap. Unless supremely well bred, their main value must lie in their stamina and resistance to current ills. It is pointless to provide them with too sheltered a life.

SAMPLE RATIONS

In designing rations for dairy cattle, it is common practice to provide their maintenance ration in the form of roughages and their production ration in the form of concentrates. This system works passably well, because, as Fig. 1 shows, the maintenance requirement of the cow is round about half of its total requirement of energy foods, and it doesn't need much protein. So the cow gets half its energy foods in the form of roughages, in which the starch equivalent forms one-third of the dry matter, and the other half in concentrates in which the starch equivalent forms two-thirds of the dry matter. In the whole ration, therefore, there is about 50 lb (22·7 kg) starch equivalent in 100 lb (45·4 kg) of dry matter; this degree of concentration is about the same as that of the better natural pastures which are the cow's natural food.

The arrangement also produces automatically a reasonably good mineral balance in the diet—half the food value being in the form of lime-rich roughages and the other half in phosphorus-rich concentrates.

The maintenance requirements of the goat are relatively small, and of her relatively large 'production' ration, a bigger proportion goes to provide the *energy* to carry out high-speed food conversion. If you design goat rations on the dairy-cow basis, you get a third of the ration in the form of roughages and two-thirds in the form of unbalanced concentrates—giving an overall ration which would be too concentrated for a cow, and much too concentrated for a goat.

As the ordinary roughages are not sufficiently rich in lime to balance twice their food value in phosphorus-rich concentrates, the resultant ration is also minerally imbalanced, in most cases. So the belief has grown up that high-yielding goats must have all their roughages and much of their fresh green food in the form of calcium-rich legume hays and fodders, supplemented by calcium-rich beet pulp and calcium-rich seaweed meal. It is indeed a fact that goats fed more or less in accordance with this arrangement will not eat any other kind of hay; it is also a fact that given this type of diet, which contains about four times as much lime as a high-yielding goat actually needs for maintenance and production, the goat is unable to breed, milk satisfactorily, or even keep alive, without a very large phosphorus-rich concentrate ration, to keep the mineral balance.

As far as quality is concerned, it is very much more satisfactory to give the goat her maintenance ration in the form of concentrates and her production ration in the form of roughages. For a 150-lb (68 kg) goat, that is $150 \times 0.9 = 1.35$ lb (612 g) starch equivalent or 2 lb (907 g) of concentrates per day. Concentrate foods and mixtures normally provide enough protein to leave a large surplus over the goat's maintenance needs to be 'carried forward' into the production ration of roughage. This ration of 2 lb of concentrates in a day is adequate and suitable for the needs of any goat giving up to 11 pints (6·2 litres) of milk a day at her peak yield; for a goat that doesn't give less than a gallon at peak yield, the same ration can be continued until her yield drops below 4 pints (2·3 litres). At this level of production it is more economical to reduce the concentrates by ½ lb (227 g) for each 1-pint (0·6 litres) fall in yield—always assuming that there is an adequate supply of fresh food and hay to keep improving condition.

The Kinloch Herd boasted a recorded herd average of 4,344 lb (1,970 kg) in 365 days on the produce of a marginal land holding in the Galloway hills. Concentrates are limited to 2 lb (907 g) per head per day, mainly in the form of home-grown bruised oats; bulk feeding consists of rough grazing, supplemented by successional crops of rape, kale and drum-head cabbage, with a patch of oats and tares to fill the gap between rape and kale. Winter feeding is based on tripoded hay from herbal leas; roots used are carrots and potatoes. The goats kid once in two years. These methods produce about 50 per cent more milk per acre than could be produced by good cows on far better ground.

The following sample diets are designed with this scheme in mind. In formulating the concentrated part of the ration, the mixtures have been made to accord in general food values with the dairy compound cakes on the market, as far as possible. So a proprietary compound can replace any of the mixtures (1), (2) and (3) without altering the balance of the whole diet.

Concentrate Mixtures

(1)   30 lb (13·6 kg) cotton-seed cake
           30 lb (13·6 kg) linseed cake
           40 lb (18·1 kg) sunflower-seed cake
         100 lb (45·4 kg) flaked maize
         100 lb (45·4 kg) whole oats
         100 lb (45·4 kg) whole rye
          40 lb (18·1 kg) seaweed meal

100 lb (45·4 kg) contains:
66·5 lb (30·2 kg) starch equivalent, 12·4 lb (5·6 kg) digestible protein, 87 lb (39·5 kg) dry matter.

2 lb (907 g) contains:
1·33 lb (598 g) starch equivalent, 0·25 lb (113 g) digestible protein, 1·7 lb (771 g) dry matter

In this mixture it is possible to ring the changes by replacing any oil-cake by any other oil-cake, and any cereal by any other cereal, provided the proportion of cereals to oil-cakes is kept. (See also the economy suggestions on p. 138.)

(2)   100 lb (45·4 kg) kibbled beans
       100 lb (45·4 kg) oats
        20 lb (9·1 kg) seaweed meal
  100 lb (45·4 kg) contains:
     60·5 lb (27·4 kg) starch equivalent; 13·3 lb. (6 kg) digestible protein; 86 lb (39 kg) dry matter.
  2 lb (907 g) contains:
     1·2 lb (544 g) starch equivalent; 0·26 lb (114 g) digestible protein; 1·7 lb (771 g) dry matter.
Any legume seed can be used to replace the beans and any cereal to replace the oats; or a mixture of pulses to replace beans and a mixture of cereals to replace oats.

(3) A winter mixture which is not very palatable, but may be rendered more so by mixing it with chopped roots or soaked beet pulp immediately before feeding. Alternatively, the constituents can be fed separately, the dried grass unground.
       100 lb (45·4 kg) oats
       100 lb (45·4 kg) grass meal
  100 lb (45·4 g) contains:
     62·1 lb (28·2 kg) starch equivalent; 12·3 lb (5·6 kg) digestible protein; 88·8 lb (40·3 kg) dry matter
  2 lb (907 g) contains:
     1·25 lb (567 g) starch equivalent; 0·25 lb (113 g) digestible protein; 1·8 lb (817 g) dry matter.
Any dried leaf may be substituted for the dried grass if the ration is increased to 2½ lb (1·1 kg); dried comfrey can be substituted without altering the quantity of the ration. Any cereal or mixture of cereals may replace the oats.

(4) A special-purpose mixture with a high mineral and vitamin content and a lower concentration, designed for feeding to dry stock, kids and any goat suffering from the effects or after effects of worm infestation.
       100 lb (45·4 kg) bran
       100 lb (45·4 kg) maize (cornflour-residue) meal (e.g. 'Paisley' meal)

30 lb (13·6 kg) seaweed meal
100 lb (45·4 kg) sugar-beet pulp (dry)
2½ lb (1·1 kg) contains:
1·44 lb (653 g) starch equivalent; 0·26 lb (114 g) digestible
protein; 2·2 lb (998 g) dry matter
Wheat-germ meal may replace the maize-residue meal. Lucerne
meal may replace the beet pulp.

Bulk diets

Bulk diets are highly dependent on the resources of the goatkeeper.
Those here suggested are intended to be typical of a range of
conditions, and for both summer and winter; the constituents of
each diet are also chosen because there is some indication available
as to their food value and mineral status. For many goat feeds there
are no scientific data available.

Bulk diet (1) is the classic winter ration of the high-yielding goat; it is
not recommended, but under some systems of management it is
inevitable; it cannot be balanced by any of the mixtures given
above, nor by any proprietary compound, unless the concentrates
are fed at a higher level than 2 lb (907 g) per day. But it can be
balanced by omitting the seaweed meal from mixtures (1) and (2)
and adding 1 per cent of fish meal—i.e. a dessertspoonful to the
2-lb ration. Otherwise it is liable to cause breeding troubles (see
p. 213).
5 lb (2·25 kg) legume hay
3 lb (1·4 kg) sugar-beet pulp
5 lb (2·25 kg) kale
It contains:
3·7 lb (1·7 kg) starch equivalent; 0·72 lb (326 g) digestible protein; 7·6 lb (3·4 kg) dry matter.
    With concentrate mixture (1) or (2) it provides:
5 lb (2·25 kg) starch equivalent; 1 lb (454 g) digestible protein;
9·3 lb (4·2 kg) dry matter.
This is adequate for a goat giving up to a gallon (4·5 litres) a day.

Bulk diet (2) is a generally applicable winter ration of hay, roots and
the proceeds of foraging for natural roughage in wintertime. As
shown here, it is deficient in lime to a small extent when used in

conjunction with concentrate mixtures (1), (2) and (4). With mixture (3), or in a hard-winter district, or if foraging time is more than an hour or so and the forage fairly plentiful, it is adequate. Otherwise the addition of an extra 3 oz (85 g) seaweed meal to the concentrates will put matters right.

    4 lb (1·8 kg) meadow hay (fair only)
    10 lb (4·5 kg) fodder beet
    2 lb (0·9 kg) heather tips

It contains:
    2·84 lb (1·3 kg) starch equivalent; 0·25 lb (113 g) digestible protein; 6·3 lb (2·8 kg) dry matter.

In conjunction with any of the concentrate mixtures, it provides for a yield of up to 6 pints (3·4 litres).

Bulk diet (3) is applicable only where the goats have extensive free range over scrub, woodland or moorland, and dispenses with hand feeding of bulk foods in winter. It works perfectly well in suitable country with two provisos: (a) that it is used in conjunction with a concentrate mixture of the type shown in (4)—that is, a rather bulky concentrate ration; (b) hay or other hand-fed roughage must be available for really stormy days. Otherwise there is the risk of chills and indigestion, when a cold goat who hasn't a good store of roughage inside her eats a quantity of concentrates. The concentrates should be fed only when the goat is full of roughage.

    5 lb (2·25 kg) tree bark
    5 lb (2·25 kg) heather tips

It contains:
    2·3 lb (1 kg) starch equivalent; 0·15 lb (68 g) digestible protein; 5·9 lb (2·7 kg) dry matter.

With the concentrate mixture, it is adequate for yields up to 4½ pints (2·6 litres).

Bulk diet (4) is the summer diet of the high-yielding goat on the high-lime-heavy concentrate system. It is very expensive, but easier to balance than the winter diet (bulk diet 1) because in summer young growing legumes contain a fair proportion of phosphorus which is lost in hay making. For this reason the minor symptoms of 'high-calcium' phosphorus deficiency (stiffness and limping due to the lack of phosphorus to release food energy into the muscles) usually clear up in summer.

20 lb (9·0 kg) lucerne
3 lb (1·4 kg) clover hay
5 lb (2·25 kg) tree leaves

It contains:

3·65 lb (1·6 kg) starch equivalent; 0·92 lb (417 g) digestible protein; 8·5 lb (3·8 kg) dry matter.

With 2 lb (907 g) of concentrate, this gives:

5 lb (2·25 kg) starch equivalent; 1·18 lb (535 g) digestible protein; 10·2 lb (4·6 kg) dry matter.

This diet is not indicative of the capacity of a heavy milker; but, as shown, it will support yields up to 9 pints (5·1 litres). It is noteworthy that, as with all summer bulk diets, there is ample protein for higher production, but not enough starch equivalent. The starch equivalent must come off the goat's back. In the winter diets, allowance was made for putting it there. Whch all goes to show that feeding in accordance with the glandular system of the goat is also feeding in accordance with the changing composition of the available fodder.

Bulk diet (5) is the typical diet of the goat on restricted range. It is to be sincerely hoped that the goat is allowed to get some such proportion of the gorse and tree leaves here listed, or their equivalent in deep-rooting and harder forage. Grass is not a balanced diet for the goat; the better the grass the worse the balance; the lack of bulk and fibre and the low mineral status relative to its digestible nutrients are equally dangerous.

20 lb (9·0 kg) pasture grass
5 lb (2·25 kg) tree leaves
5 lb (2·25 kg) gorse

It contains:

3·84 lb (1·7 kg) starch equivalent; 1·07 lb (485 g) digestible protein; 7·6 lb (3·4 kg) dry matter.

With the concentrate ration these provide for 10 pints (5·7 litres) of milk, and leave some room for more in a good-sized goat.

Bulk diet (6) is appropriate to the goats on extensive free range—not to be confused with those sharing a fair-sized paddock with a Shetland pony, who are catered for in (5) above. It is the cheapest diet here—it should touch the heart of the hard-bitten dairy farmer, the thought of turning the following rubbish into milk:

15 lb (6·8 kg) tree leaves
15 lb (6·8 kg) 'weeds' (average of chicory, dandelions and nettles)
5 lb (2·25 kg) grass
5 lb (2·25 kg) brushwood

It contains:

4·05 lb (1·8 kg) starch equivalent; 0·81 lb (367 g) digestible protein; 10·25 lb (4·6 kg) dry matter.

This will make 7 pints (3·9 litres) of milk, without any concentrates. Thousands of goats either can eat more of this sort of diet or take a more favourable view of its nutrient value; they certainly produce more than 7 pints on it.

The goat's maintenance and a gallon of milk require roughly:

For maintenance	*Production*	*Total*
1·36 lb (0·6 kg) starch equivalent	3·25 lb (1·5 kg) starch equivalent	4·61 lb (2·1 kg) starch equivalent
0·13 lb (59 g) digestible protein	0·5 lb (227 g) digestible protein	0·73 lb (331 g) digestible protein

In calculating the milk yields supported by a diet, the milk yield supported by the digestible protein of the diet has been, in general, given. On typical autumn and winter foods this leaves a surplus of 1 lb (454 g) or more of starch equivalent per day to go on the goat's back. On typical spring and summer foods it takes about 5 oz (142 g) of starch equivalent a day off the goat's back, that is about 1¼ oz (25 g) of fat. Where the diet, e.g. legumes, is naturally unbalanced for the goat, yields have been stated more conservatively.

No mention has been made of a diet for stall feeding; the variety of materials available for the purpose is so great as to make a 'typical' stall-feeding or yarding diet unattainable. But here follows a table of the nutrient value and dry-matter content of about 150 goat foods.

The table has been compiled solely for the use of goatkeepers, and not for reference for any other purpose. Faced by the fact that for only a small proportion of goat foods was a chemical analysis available, and that only for a still smaller proportion were the results of digestibility trials available, I have worked to the principle that an intelligent guess is more helpful than a stack of undigested figures. All the foods mentioned here have been chemically analysed;

where the results of digestibility trials are available these have been used; otherwise I have estimated the digestible nutrients from the chemical constituents on the basis of normal theoretical expectations, backed by a knowledge of what goats eat and how much they like it. For instance, digestibility trials have been carried out on poplar leaves but not on oak leaves: they are similar in starch-equivalent: fibre ratio, and I have estimated the digestibility of oak leaves from that shown for poplar leaves. Tree bark has not been subjected to digestibility trials, but brushwood has. I have assumed that the carbohydrates and fibre found in the chemical analysis of tree bark are digested in the same proportions as those of brushwood have been shown to be. So tree bark turns up with a starch equivalent better than meadow hay—which is certainly no surprise to goats, whoever else may be dubious.

The layout is designed for quick practical reference.

Build your ration on the digestible protein you need; build it of seasonable bulky foods for the most part, and you will find that the starch equivalents accord with the needs of the season, giving you a surplus in winter and a small deficiency in summer. The goat's digestion works best when the starch equivalent is rather less than half the dry matter; it must not be allowed to become much more than half. Dry-matter capacity varies: 5 to 7 lb (2·25 to 3·2 kg) a day, according to size of goat, is near the minimum for health and comfort. Allow a big goat with a big yield up to 10½ lb (4·7 kg) dry matter without question—she will probably take 12 lb (5·4 kg) if the food is appetizing. On winter feeding 9 lb (4·1 kg) is plenty for most goats, unless the fodder is of superlative quality. It is worth weighing your goats. Weights vary between 90 and 250 lb (40·8 and 113·4 kg); appetites and maintenance needs vary with them; to feed and breed progressively, it helps to know the goat's capacity in relation to her bodyweight. (See Appendix 4).

Table 5: Nutrient value and dry-matter content of goat foods

100 lb (45·4 kg) of	Starch equivalent lb	Starch equivalent kg	Contains: Digestible protein lb	Contains: Digestible protein kg	Dry matter lb	Dry matter kg	Mineral state	Remarks
Tree leaves								
Ash	14·1	6·4	1·6	0·7	20·0	9·1	High lime	Very palatable, laxative
Beech	2·1	0·95	0·47	0·2	20·0	9·1	Med. lime	Acceptable
Elm	11·4	5·2	2·6	1·2	20·0	9·1	High lime	Very palatable
Horse chestnut	8·7	3·9	2·1	0·9	20·0	9·1	High lime	Palatable
Mixed leaves (July)	9·0	4·1	1·47	0·7	20·0	9·1	High lime	Palatable
Oak	8·5	3·8	3·5	1·6	20·0	9·1	Med. lime	Binding, palatable
Poplar	8·2	3·7	1·4	0·6	20·0	9·1	High lime	Acceptable
Willow	7·6	3·4	1·9	0·9	20·0	9·1	High lime	Palatable
Leaves of non-legume plants								
Artichoke tops	16·2	7·3	2·0	0·9	32·0	14·5	Med. lime	Palatable, fresh
Beet tops	8·6	3·9	1·4	0·6	16·2	7·3	Med. lime	Palatable, withered
Cabbage								
drumhead	6·6	3·0	1·1	0·5	11·0	5·0	Adequate	Acceptable
open-leaved	9·5	4·3	1·8	0·8	15·3	6·9	Adequate	Palatable
Carrot tops	7·7	3·5	2·21	1·0	18·2	8·2	Med. lime	Acceptable

100 lb (45·4 kg) of	Starch equivalent		Contains: Digestible protein		Dry matter		Mineral state	Remarks
	lb	kg	lb	kg	lb	kg		
Chicory	5·0	2·3	0·66	0·3	11·5	5·2	Adequate	Palatable
Comfrey, Russian	5·32	2·4	2·04	0·9	12·4	5·6	High lime	Palatable
Dandelion	7·8	3·5	1·7	0·8	14·5	6·6	High lime	Palatable
Hops	8·1	3·7	2·1	0·9	23·0	10·4	High lime	Palatable
Kale								
thousand-headed	10·3	4·7	1·7	0·8	15·8	7·1	Adequate	Palatable
marrow-stem	9·1	4·1	1·7	0·8	14·0	6·3	Med. lime	Palatable
Mangold tops	5·3	2·4	1·7	0·7	11·7	5·0	Med. lime	Palatable
Nettles	12·6	5·2	3·5	1·6	24·0	10·9	Very high lime	Palatable in seeding stage, always acceptable
Potato haulm	7·2	3·3	1·1	0·5	23·0	10·4	Med. lime	Palatable
Turnip tops	5·3	2·4	1·5	0·7	11·6	5·3	Low	Acceptable, clean
Fresh legumes in first flower								
Alfalfa(see Lucerne)								
Alsike	6·3	2·8	2·1	0·9	15·0	6·8	High lime	Palatable
Beans	7·1	3·2	2·3	1·0	15·0	6·8	High lime	Palatable
Clover								
American crimson	7·4	3·3	4·2	1·9	19·0	8·6	High lime	Palatable

crimson	8·9	4·0	2·1	0·9	18·5	8·4	High lime	Palatable
red	10·2	4·6	2·5	1·1	19·0	8·6	High lime	Palatable
white	8·8	4·0	2·8	1·3	18·5	8·4	High lime	Palatable
Kidney vetch	7·9	3·6	1·4	0·6	18·0	8·2	Med. lime	Palatable
Lucerne	10·3	4·7	3·1	1·4	24·0	10·9	High lime	Palatable
Peas	6·8	3·1	2·4	1·1	16·8	7·6	Med. lime	Palatable
Sainfoin	7·6	3·4	2·3	1·0	20·0	9·1	Med. lime	Palatable
Trefoil	9·1	4·1	2·4	1·1	20·0	9·1	High lime	Palatable
Vetches	7·5	3·4	2·2	1·0	17·5	7·9	Med. lime	Palatable

Flowering cereals and grasses

Barley	16·1	7·3	1·5	0·7	31·4	14·2	Lime def.	Palatable
Buckwheat	8·1	3·7	1·6	0·7	16·3	7·4	Lime def.	Palatable
Maize	9·1	4·1	1·0	0·4	19·4	8·8	Lime def.	Palatable
Millet	5·4	2·4	0·7	0·3	13·0	5·9	Lime def.	Palatable
Oats	10·0	4·5	1·4	0·6	23·2	10·5	Lime def.	Palatable
Rye	11·3	5·1	2·1	0·9	23·4	10·6	Lime def.	Palatable
Ryegrass								
perennial	10·6	4·8	1·8	0·8	24·8	11·2	Low	Acceptable
Italian	11·4	5·2	2·1	0·9	25·0	11·3	Low	Acceptable
Timothy	13·0	5·9	1·5	0·7	30·0	13·6	Low	Acceptable
Pasture grass								
close-grazed	14·7	6·7	4·5	2·0	20·0	9·1	Low	Acceptable if clean
rotationally grazed	14·6	6·6	3·7	1·7	20·0	9·1	Low	Acceptable
extensive grazing	11·2	5·1	2·5	1·1	20·0	9·1	Low	Acceptable, clean

100 lb (45·4 kg) of	Starch equivalent lb	kg	Contains: Digestible protein lb	kg	Dry matter lb	kg	Mineral state	Remarks
Roots								
Artichokes	16·4	7·4	1·0	0·4	20·4	10·3	Low	Acceptable, laxative
Carrots	8·8	4·0	0·8	0·4	13·0	5·9	Low	Acceptable
Fodder beet	13·0	5·9	0·8	0·4	19·3	8·7	Low	Palatable
Kohlrabi	8·3	3·8	0·7	0·3	12·7	5·8	Low	Acceptable
Mangolds intermediate	6·2	2·8	0·7	0·3	12·0	5·4	Low	Palatable, not males
Parsnips	10·6	4·8	1·0	0·4	15·0	6·8	Low	Acceptable, toxic?
Potatoes	18·5	8·4	1·1	0·5	23·8	10·8	Lime def.	Palatable
Swedes	7·3	3·3	1·1	0·5	11·5	5·2	Low	Acceptable
Turnips	4·4	2·0	0·6	0·3	8·5	3·8		Edible
Silage								
Grass	12·6	5·7	2·4	1·1	24·5	11·1	Low	Acquired taste
Lucerne	11·1	5·0	2·8	1·3	25·0	11·3	High lime	Acquired taste
Maize	12·1	5·5	1·5	0·7	21·0	9·5	Lime def.	Acquired taste
Oats	8·9	4·0	1·1	0·5	23·7	10·7	Lime def.	Acquired taste

Pea haulms and pods, cannery waste	11·6	5·3	2·0	0·9	23·6	10·7	High lime	Acquired taste
Rye	5·5	2·5	0·9	0·4	13·1	5·9	Lime def.	(Bad silage)
Vetch and oats	13·0	5·9	3·8	1·7	34·6	15·7	Adequate	Acquired taste

Natural roughage

Brushwood	14·8	6·7	2·1	0·9	75·0	34·0	Low	Palatable
Gorse	8·9	4·0	2·2	1·0	51·0	23·1	Adequate	Young gorse tips are palatable and more nutritious
Heather tips	12·2	5·5	1·4	0·6	50·0	22·7	Med. lime	Palatable in winter
Tree bark sweet chestnut	34·0	15·4	1·6	0·7	67·5	30·6	Low	Palatable, also willow, holly, elm and fruit tree

Hays

Barley	42·7	19·4	4·6	2·1	86·0	39·0	Low	Palatable
Clover crimson	33·0	14·9	8·3	3·8	83·0	37·6	High lime	Palatable
red	38·0	17·2	8·5	3·8	83·0	37·6	High lime	Palatable
Couch grass	65·0	29·5	4·7	2·1	93·0	42·2	Low	?
Lucerne	27·0	12·2	9·7	4·4	83·0	37·6	High lime	Palatable

100 lb (45.4 kg) of	Starch equivalent		Contains: Digestible protein		Dry matter		Mineral state	Remarks
	lb	kg	lb	kg	lb	kg		
Meadow								
poor	22.0	10.0	3.4	1.5	86.0	39.0	Low	Edible in part
medium	32.0	14.5	3.6	1.6	85.0	38.5	Low	Acceptable
good	37.0	16.8	5.4	2.4	86.0	39.0	Low	Palatable
Moorland hay molinia and bent	24.6	11.1	5.5	2.5	86.0	39.0	Deficient	Acceptable
Oats	34.0	15.4	4.4	2.0	86.0	39.0	Low	Palatable
Rye	44.0	19.9	7.3	3.3	86.0	39.0	Low	Palatable?
Ryegrass perennial	38.0	17.2	6.4	2.9	86.0	39.0	Low	Palatable if leafy
Italian	44.0	19.9	6.9	3.1	86.0	39.0	Low	Wastefully used
Ryegrass and clover	30.0	13.6	6.2	2.8	86.0	39.0	Med. lime	Very wastefully used
Sainfoin	40.0	18.1	6.2	2.8	86.0	39.0	High lime	Palatable
Timothy	35.0	15.9	4.0	1.8	86.0	39.0	Low	Edible
Trefoil	37.0	16.8	11.8	5.3	84.0	38.1	Med. lime	Palatable
Vetches	32.0	14.5	9.4	4.3	84.0	38.1	High lime	Palatable
Vetches and oats	34.0	15.4	6.5	2.9	84.0	38.1	Adequate	Palatable
Wheat	40.6	18.4	3.1	1.4	86.0	39.0	Low	Acceptable

Dried leaves

Artichoke	37.3	16.9	7.6	3.4	87.5	39.7	Adequate	Palatable
Ash	62.0	28.1	7.0	3.2	88.0	39.9	High lime	Palatable (very)
Beech	8.9	4.0	2.0	0.9	84.6	38.4	Med. lime	Edible
Chicory	40.0	18.1	5.3	2.4	93.8	42.5	Adequate	Palatable
Comfrey	40.9	18.5	13.0	5.9	88.9	40.3	High lime	Palatable
Elm	50.0	22.7	11.6	5.3	88.0	39.9	High lime	Palatable (very)
Grass	65.7	29.8	16.7	7.6	90.0	40.8	Adequate	Palatable, unground
Horse chestnut	37.5	17.0	9.0	4.1	86.0	39.0	High lime	Palatable
Lucerne	44.1	20.0	11.6	5.3	91.0	41.3	High lime	Palatable
Nettles	46.5	21.1	12.8	5.8	88.6	40.2	High lime	Palatable
Oak	36.0	16.3	16.0	7.3	85.2	38.6	Med. lime	Palatable
Poplar	34.0	15.4	6.0	2.7	84.0	38.1	High lime	Acceptable
Willow	34.0	15.4	8.5	3.8	89.0	40.4	Hign lime	Palatable

Straws and chaffs

Barley straw	23.0	10.4	1.8	0.8	86.0	39.0	Low	Edible
Bean straw	19.0	8.6	2.2	1.0	86.0	39.0	High lime	Acceptable
Lupin, sweet blue	25.0	11.3	2.4	1.1	87.3	39.6	High lime	Acceptable
Oat chaff	20.0	9.1	1.0	0.4	86.0	39.0	Low	Acceptable
Oat straw	20.0	9.1	1.7	0.8	86.0	39.0	Low	Acceptable
Pea chaff	64.2	29.1	3.6	1.6	88.0	39.9	High lime	Acceptable
Pea straw	17.0	7.7	4.3	1.9	86.0	39.0	Low	Acceptable

100 lb (45·4 kg) of	Starch equivalent lb	kg	Contains: Digestible protein lb	kg	Dry matter lb	kg	Mineral state	Remarks
Fruits								
Acorns	41·2	18·7	2·7	1·2	50·0	22·7	Low	Acceptable ripe
Apples	12·4	5·6	1·7	0·8	26·0	11·8	Low	Palatable
Ash fruits, dry	70·0	31·7	6·5	2·9	85·0	38·5	High phosphorus	Palatable
Chestnut								
horse, shelled	38·0	17·2	2·6	1·2	50·8	23·0	Low	Acceptable
sweet	64·0	29·0	6·0	2·7	60·0	27·2	Low	Palatable
Cleavers	49·0	22·2	8·5	3·8	87·0	39·5	?	Palatable
Elm fruits (dry)	50·0	22·7	18·0	8·2	88·0	39·9	High phos.	Palatable
Hawthorn	40·0	18·1	0·7	0·3	30·0	13·6	Low	Palatable
Rose hips	14·0	6·3	1·0	0·4	28·0	12·7	High lime	Palatable
Cereal grains								
Barley	71·4	32·4	7·6	3·4	85·0	38·5	High phos.	Palatable
Maize	77·6	35·2	7·9	3·6	87·0	39·5	High phos.	Palatable
Oats	59·5	27·0	8·0	3·6	86·7	39·3	High phos.	Palatable
Rye	71·6	32·5	9·6	4·3	86·6	39·3	High phos.	Palatable
Wheat	71·6	32·5	10·2	4·6	86·6	39·3	High phos.	Palatable

Legume seeds

Beans	65·8	29·8	20·1	9·1	85·7	38·9	High phos.	Palatable
Lupins								
blue	72·1	32·7	30·3	13·7	87·3	39·6	High phos.	Acceptable
yellow	71·1	32·3	38·1	17·3	87·7	39·8	High phos.	Acceptable
Peas	69·0	31·3	19·4	8·8	86·0	39·0	High phos.	Acceptable

Oil seeds

Linseed	119·0	54·0	19·4	8·8	93·0	42·2	High phos.	Palatable
Sunflower	103·0	46·7	12·8	5·3	92·5	41·9	High phos.	Acceptable

Oil cakes

Cotton seed	41·6	18·9	17·8	8·1	87·9	39·9	High phos.	Palatable
Groundnut	77·5	35·1	42·0	19·0	89·7	40·7	High phos.	Palatable
Linseed	74·0	33·6	25·0	11·3	88·8	40·3	High phos.	Palatable
Palm kernel	73·2	23·2	17·5	7·9	89·0	40·4	High phos.	Palatable
Sesamum	73·0	33·1	40·0	18·1	90·7	41·1	High phos.	Palatable
Soya bean	68·9	31·2	38·8	17·6	85·5	38·8	High phos.	Palatable
Sunflower	72·5	32·9	33·6	15·2	90·5	41·0	High phos.	Palatable

Germ meal and brans

Linseed bran								
flax chaff	29·8	13·5	4·8	2·2	87·3	39·6	High phos.	Edible
Maize bran	67·0	30·4	5·5	2·5	88·2	40·0	High phos.	Palatable
Maize-germ meal	84·3	38·2	10·4	4·7	89·2	40·5	High phos.	Palatable
Maize-gluten feed	75·6	34·3	20·0	9·1	89·6	40·6	High phos.	Palatable

100 lb (45·4 kg) of	Starch equivalent		Contains: Digestible protein		Dry matter		Mineral state	Remarks
	lb	kg	lb	kg	lb	kg		
Maize (cornflour-residue) meal	72·6	32·9	17·0	7·7	89·8	40·7	High phos.	Palatable
Pea-husk meal	64·2	29·1	3·6	1·6	87·8	39·8	Low	Palatable
Pea-pod meal, cannery waste	54·8	24·9	9·7	4·4	90·0	40·8	High phos.	Palatable
Wheatfeed (75 per cent extraction)	49·0	22·2	12·2	5·5	87·0	39·5	High phos.	Acceptable
Wheat-germ meal	54·0	24·5	20·0	9·1	90·0	40·8	High phos.	Palatable
Bran	42·0	19·0	10·9	4·9	87·0	39·5	High phos.	Acceptable
Miscellaneous								
Brewers' grains (dry)	48·3	21·9	13·0	5·9	89·7	40·7	High phos.	Acceptable
Distillers' grains (wet)	16·2	7·3	6·2	2·8	26·2	11·9	High phos.	Palatable
Milk								
skim	8·1	3·7	3·0	1·4	9·2	4·2	Adequate	Palatable
whole	18·1	8·2	3·0	1·4	12·8	5·8	Adequate	Palatable
Seaweed meals								
Fucus	37·0	16·8	–	–	88·7	40·2	High lime	Palatable
Laminaria	38·7	17·5	6·1	2·8	83·7	38·0	High lime	Palatable
Sugar-beet pulp	60·6	27·5	5·3	2·4	90·0	40·8	High lime	Palatable

A note on seaweed meals

The analysis of any brand of seaweed meal varies with the types of seaweed composing it and the time and place of harvesting. Three main types of seaweed are used: knobbed wrack (the most commonly used), tangle and bladderwrack. A full typical analysis of each is given below, along with the advertized analysis of a 'Natural Mineral Mixture', based on seaweed meal, but utterly different from them in mineral balance. References to seaweed meal in this book do not include such mixtures.

Composition per cent	Knobbed wrack (*Ascophyllum*)	Tangle (*Laminaria*)	Bladderwrack (*Fucus*)	'Natural Mineral Mixture'
Digestible protein	6·82	6·1	0	9·82
Carbohydrates	58·09	45·8	53·6	56·08
Fibre	4·5	8·6	9·4	4·2
Ether extract	2·26	1·1	4·2	2·4
Ash	18·85	16·8	16·3	16·82
Moisture	9·48	16·3	11·3	10·68
Sand	1·48	?	?	1·02
Calcium	1·49	4·62	1·60	1·29
Phosphorus	0·12	0·21	0·11	0·82 (n.b.)
Magnesium	0·71	?	?	1·30
Sodium	1·97	?	?	1·17
Potassium	2·50	1·25	1·80	2·87
Chlorine	1·11	0·69	1·32	0·91
Iodine	0·07	?	?	0·16
Cobalt	trace	trace	trace	0·11

Chapter Eight

Selection of Breeding Stock

This chapter is intended to serve the needs of the person who wishes to produce goats' milk in quantities, great or small, and not those of the pedigree breeder whose primary purpose is to produce stock for sale. Consequently the science of genetics is only referred to in the broadest terms, and the available space is devoted to the selection and breeding management of stock suited to specific methods of goat dairying.

The dairy farmer's breeding policy is to provide a regular flow of milk, and an adequate supply of satisfactory young stock for herd replacement with the minimum of cost and risk. In selecting his breeding stock he must always play safe, and resist the lure of gambling in genetics. In particular he must avoid as far as possible the dangers of inbreeding, and at the same time say farewell to his hopes of achieving perfection in his herd. For the pedigree breeder, a knowledge of genetics is useful; and an intimate knowledge of the strains of the breed in which he is working is essential. Goat breeds are of such recent origin that the only breed characteristics which are established with any degree of uniformity are those affecting colour and appearance; inbreeding is therefore essential to the pedigree breeder who wishes to produce a strain of distinctive character and true-breeding potentialities. The pedigree breeder

Plate 10. SM CH. §§44/36†Mostyn Minimayson BR. CH. Sire of seven full champions and many other prominent prize winners and milkers. Six times winner of Stud Goat Cup. So indelibly has he stamped his beautiful quality and conformation on his daughters and passed on the great milking ability of his forebears, that breeders from far and wide have travelled their goats for his services. Minimayson must unquestionably be termed 'The Greatest'. S. Breaston Byrnian BR. CH. D. RM44 Mostyn Minimay Q*1.

must not only take the risk of inbreeding to fix desired character, but must proceed to the even more expensive business of proving and publicizing the character produced; he must design his feeding and management to ensure the maximum possible production from his goats, have their yields officially recorded and their confirmation officially approved in the show ring. Maximum possible production is never the most economic production; the pedigree breeder's overheads will usually ensure that the milk of his herd is expensive to produce. One of the main handicaps to the development of a goat-dairying industry in Britain is that far too many goat owners accept the expensive responsibility of the pedigree breeder, and far too many pedigree breeders lack the stock or personal qualities to justify their status. There is a dearth of first-rate goat-dairy farmers, and a surplus of second-rate stock breeders.

The main considerations in selecting stock for milk production are their inherent milking capacity, food-intake capacity and food-to-milk conversion efficiency. The pedigree breeder generally aims to produce an animal with all these qualities developed as highly as possible. But the degree to which the dairy farmer needs each of these distinct qualities will depend upon the system under which his goats are managed. While it is not possible to discuss the special needs of every conceivable method of management, useful distinction can be made between three main types of management: i.e. *stall feeding* or *yarding*, under which system the goats have most of their food cut and carried to them; *goat farming on improved land*, in which case the goats obtain the bulk of their ration grazing crops which have been specially grown for them and other farm stock; *free range on scrub and rough grazings*, in which case the goats obtain the bulk of their ration foraging growth which would otherwise be wasted. Each of these systems of management calls for a different set of qualities in the goat who is to make the most of them.

The labour of feeding the stall-fed goat is generally the biggest item on the cost sheet, with the price of the food running it a close second. The type of goat needed is one which will produce the maximum of milk for the minimum of food and attendance. The goat's milking capacity should be high, and her efficiency in converting food to milk of the very best; but there is no call here for a big food-intake capacity. The modern dairy goat has been developed to consume a prodigious daily ration of cheap roughage and convert it into milk. But roughage which is dirt-cheap on the ground where it grows becomes progressively more expensive every time it is handled; and the dearest food stuff, per pound of digestible protein, fed to farm stock.

One solution to the economic problem set by the appetite of the modern dairy goat is to feed it large quantities, not of roughages but of highly digestible succulents—cultivated fodder crops—and a generous measure of concentrates. This system is as good as any for forcing yields to the maximum for official recording, and is popular with pedigree breeders. For the domestic goatkeeper with a few goats, who can grow most of the special fodder crops in the garden in his or her spare time, it is also suitable. But the high cost of field cultivation and handling of bulky and sappy crops is handicap to commercial dairying.

In countries such as Italy and Spain, where commercial dairying with stall-fed goats is widely practised, the type of goat used is a small, flat-sided, milky creature such as the Maltese or Malaga goat (which are probably derived from the wild *Capra prisca*). These goats are fed on industrial by-products such as tomato-cannery waste and olive pulp, plus a little hay and a small concentrate ration. They consume little more than half as much food in proportion to bodyweight as our leading breeds, but produce almost as much milk in proportion to bodyweight as our 200-gallon (909 litres) milkers. These *Capra prisca* derivatives seem adjusted to a more concentrated ration than is healthy for goats derived from the mountain goat *Capra aegagrus*. In Italy and Spain they are preferred to the available alternative type with bigger food capacity and higher yields per head, because they save labour and are more economic, even at comparatively low Spanish wage rates.

There are several British industries, at present pouring their edible wastes into rivers, pig troughs and sewage pipes, which might support stall-fed goat dairies. Though they do not appear in the show ring or the leading pedigree herds, there are a number of flat-sided goats with small food capacity and good milking abilities. The true *Capra prisca* type, with a short, straight, twisted horn, are rare.

Whether we adopt the fodder crop or industrial waste as a basis of feeding for stall-fed or yarded goats, we are adopting a very *wet* diet; this is an essential to maximum production from a given quantity of nutrients. The goat which will make the best use of such a diet is not the one who has the biggest capacity for bulk, but the one with maximum capacity for slop—that is, a goat that is deep but not wide in the body.

It is also interesting to note that the goats used for commercial dairying on the stall-feeding system in the Mediterranean basin are long-haired. On the other hand, the free-range foragers, even in mountainous districts, are preferred short-coated, like the Murcian and Granada goats of Spain. Goats derive their heat mainly from the bacterial fermentation of roughage, and on a concentrated diet are more subject to chills, even when housed, than when full of fibrous forage on the bleakest mountain top. Our stall-fed goats are often insulated from the cold by flannel coats or a layer of sub-cutaneous fat; flannel coats are less foolproof and fat is more expensive than long hair, which can always be clipped away from the udder for the sake of dairy hygiene.

The goat farmer on improved land has need of quite another kind of goat. His labour costs are largely concerned in the organization of controlled grazing; his food costs are moderate. The fewer goats he has to control, the better; the more forage crop they will eat, the less they will need out of the bag of expensive concentrates. He needs goats with as high a milking potential as possible, and a good food-intake capacity. A very high food-to-milk conversion efficiency is not always desirable; fodder crops are succulent rather than fibrous, and the sudden chills of the British climate are notorious. Goats grazing improved land will resist chilling and feed in worse weather if they carry a little fat or a coat of long hair (the flannel substitute is unsatisfactory out of doors, especially in the rain). The weight of a big udder, a well-filled and capacious paunch and a rather fat body are most easily carried on short, sturdy legs. Ground clearance for the udder is of little account in cultivated fields, but the udder requires to be exceptionally well hung, broad in the base and close to the body to avoid the dangers of chilling.

The leading breeders produce excellent stock for this purpose.

Goats that have free range on rough grazings are consuming food that costs next to nothing; the more they eat of it and the more efficiently they convert it to milk the better. Big food capacity is of prime importance. Such goats consume ample fibre to keep themselves warm and have no need of a quilt of fat or long hair, so their food-to-milk conversion efficiency may be very high. But very great milking capacity is not wanted. Any advantage to be gained from a superlative yield is more than cancelled out by the large proportion of concentrate necessary to maintain it, the burden of carrying an unwieldy udder over rough ground, and the damage that udder will suffer from the brambles and hazards of a typical rough grazing. Long legs are needed to give the udder ground clearance, and are an important asset to a forager among trees and shrubs. The quality and hang of the udder is not a matter of primary importance. As a free-range goat farmer, the writer has happiest memories of some tough, pendulous bags, with skin like crêpe rubber, impervious alike to knocks, scratches and climate. The classic peach-skinned vessel, delectable to handle and easy on the eye, is so firmly attached that it is the loser in every encounter; after a day on range its tenderness may make more difficulties for the milker than a more unsightly and tougher bag.

The amount of damage that the udder of a heavy-milking goat receives, while on the range among scrub and rock, depends to some extent on the general conformation of the goat. This matter has received less attention than it deserves from our breeders. Viewed from the side, most of our goats show an almost horizontal top line, from the shoulders to the rump; if anything the shoulders are a little

(a)   (b)

Fig. 14. Conformation of scrubland and mountain goats
(a) The Murcian goat has the scrubland shape.
(b) The Toggenburg goat has the mountain shape

higher than the top of the rump; the under line slopes steeply down from the chest to the udder. This conformation gives the goat, in appearance and reality, perfect balance and control on steep slopes; it is typical of mountain goats the world over, and has been impressed on our stock by the influence of the Swiss breeds. Goats adapted to scrubland, as opposed to mountain, pastures have a different shape; the top line slopes quite steeply up from the shoulders to the top of the rump; and the under line is roughly horizontal from the top of the udder to the chest. This conformation gives better ground clearance to the udder; the weight of the udder is more equitably shared between fore and hind legs; and the extra-long hind legs give the goat a higher reach when browsing. It is exemplified in the heavy-milking Murcian goat of Spain, in the little scrub goats of Galicia and in many of the Oriental breeds. Of home-made breeds, the Anglo-Nubian possesses this character, a legacy from its desert ancestors, and the British Alpine shows the character to a rather smaller extent.

Where goats have a very extensive range, it is a matter of importance that their herdsman should be able to see them at a distance. The camouflage markings of the Toggenburg and British Alpine are a negation of utility in this respect, and the nondescript colouring of the Anglo-Nubian is little better. Pure white or unbroken black are the colours most easily spotted at a distance. Those who doubt the visibility of black should ask a hill shepherd the colour of his dog.

There are eight recognized breeds of goat in Britain.*

(1) The Saanen

Nominally of Swiss origin, goats of this breed are all descended from imported stock, the majority of which came from the flat fields of Holland. White in colour, placid in disposition, rather short in the leg and capable of the very highest yields, this breed is well suited to goat farming on improved land. The labour involved in avoiding unsightly staining of the white coat is a disadvantage if the breed is kept under intensive conditions, where the goats spend much of their time lying in their own droppings and urine. White-skinned Saanens develop skin cancer in sunny climates; dark-skinned, but white-haired, goats are immune. Short legs and big udders associate badly with the brambles and hazards of a rough grazing. Udders in this breed are usually shapely and well hung. Butter-fat percentages are good—around 4 per cent.

(2) The British Saanen

Of mixed origins this is usually a considerably heavier and slightly leggier goat than the pure Saanen, but otherwise similar. Udders are

* Plus the old English and the Nubian, recognized but extinct.

less shapely on the whole; but yields as high as any. Like the Saanen, and for similar reasons, it is best adapted to goat farming on improved land, where the tendency of some strains to run to fat does not go amiss.

Plate 11. Miss Mostyn Owen with a prize-winning group. CH:RM49 Mostyn Daphne Q*8 BR.CH., lifetime yield 38,516 lbs – a British record; CH:R40 Mostyn Marjenka Q*1; Mostyn Marypoppin, goatling daughter of Marjenka, subsequently CH:RM46 Q*1.

(3) The Toggenburg

Descended from imported Swiss stock, this is numerically the weakest of the recognized breeds; and progress with the breed in this country is hampered by lack of numbers and inbreeding. Brown and white in colour, with an active but affectionate disposition, the Toggenburg is a small goat, usually under 100 lb (45·4 kg) adult weight. Yields are low, seldom exceeding 200 gallons (909 litres) per annum, with the butter-fat percentages of well-fed specimens deplorably low at under 3 per cent. Udders are usually well hung. Not every household requires 200 gallons of milk a year. As a stall-fed household goat, the Toggenburg has the advantage of requiring less food than most, and its brown coat resists staining. As a free-range rough grazer the Toggenburg is in its traditional environment, and on a lower level of nutrition the butter-fat content of its milk is not so outstandingly bad. There is a good demand for pedigree stock for export.

Plate 12.   The first ever Toggenburg champion, R142 Spean Meliflous Q*BR. CH., bred and owned by Mrs. J. Shields. Champion at nine shows in 1979.

(4) The British Toggenburg

Of mixed origin, this breed carries a large proportion of Toggenburg blood, which is needed to fix the rather elusive Toggenburg colour and markings. But it is a big ten-stone goat, giving notably high yields of milk with a butter-fat percentage of around 3·5. The high metabolic rate necessary for heavy production, superimposed on the warm-hearted Toggenburg disposition, tend to make the breed remarkably excitable. Though this may be a slight disadvantage in a strictly commercial herd, it makes the 'B.T.' the most responsive pet in the goat world. The relatively stain-proof coat is an added advantage under intensive conditions. Long legs, high-geared temperament and big food capacity suit the breed for free range on rough grazings, but badly hung udders are a common fault. As the B.T. is restrained with more difficulty than other breeds, and its rather low butter fat is further lowered by soft, rich feeding, it is not the obvious choice for goat dairying on improved land.

(5) The British Alpine

This shares much ancestry with the British Toggenburg, but is black with white Swiss markings; a big, leggy goat, turning the scale around 140 lb (64 kg), it is capable of high yields, with a butter-fat percentage of about 4 per cent. It is the only British-made breed to have been developed mainly under free-range conditions, and has an independence of character which would be better appreciated if there were more men farming goats, and fewer women keeping them as pets. Under intensive conditions, the B.A. is almost as easy to keep clean and as efficient a producer as the British Toggenburg, but a duller companion. For arable dairying, it is rather less easy to control than the Saanen or British Saanen, and is relatively free from the tendency to run to fat; but unwieldy udders are not uncommon in the breed, and give rise to trouble under these conditions. The black coat attracts flies, which is a handicap in closely wooded country. It remains the most adaptable all-round breed, and could be developed into a first-rate goat for scrub dairying with less difficulty than any other.

(6) The Anglo-Nubian

Of mixed origin, this breed owes its distinctive features to imported goats of India Jumna Pari and Egyptian Zariby type. Undoubtedly the most important British contribution to world goat breeding, the Anglo-Nubian is the most distinctive of our recognized breeds in both appearance and performance. Colours are various, with roan and white predominating; lop ears and Roman nose are typical. It is more heavily fleshed than the Swiss breeds, weighing about 150 lb (68 kg), gives yields comparable with the Toggenburg in quantity (up to 300 gallons [1,364 litres] a year), but with a butter-fat percentage of around 5 and a percentage of solids-not-fat of 10 to 11, compared with the 8 to 9 of other breeds. The rich milk characteristics of the flat-sided goats of the sub-tropics are here blended with the great fodder capacity of the Swiss breeds to make an economic producer of what is perhaps the most highly digestible and perfectly balanced food available to mankind.

The udder of the Anglo-Nubian, though not remarkable for its shapeliness, has a better ground clearance than that prevailing in other breeds. Though the coat is short, beneath it lies an insulating layer of subcutaneous fat which enables the goat to withstand adverse weather as well as most other breeds. It is excellently suited to free-range goat farming, or to arable dairying, under which conditions it can produce milk *solids* as economically as any. But being an inefficient producer of milk *water,* unless an exceptionally discerning human consumption market will pay a premium for its quality, Anglo-Nubian milk is best cashed when fed to stock. As a house goat the Anglo-Nubian has the disadvantage of being exceptionally vocal; but its milk is undoubtedly the most desirable of any.

(7) The Golden Guernsey

A survival from the Channel Islands, which owes its existence to Miss Miriam Milbourne and her painstaking breeding programme. The first of these goats were imported from Guernsey in 1965 and

Plate 13. Novington Boris GG 85H. Winner of many prizes at open shows including Best Adult Male and Reserve Best Exhibit, Devon Goat Society Show; Best A.O.V. Male, Warwickshire Goat Society Show; 3rd Sire and Progeny, Cornwall Male and Young Stock Show. Breeder Miss R. E. Carney, owner Mrs M. Rosenberg.

two years later the English Golden Guernsey Goat Club was founded. British Goat Society recognition of the breed followed in 1970 with the opening of a Golden Guernsey Register. An English Guernsey section followed in 1974, building on the first cross of G.G. female to S. or B.S. male.

Ideally, a Golden Guernsey should have erect ears, slightly curled at the tip, straight or slightly dished facial outline and be horned, hornless or disbudded. Its colour should be golden as should its skin, and its coat should be short, with or without long hair on spine and quarters. It should be light-boned, of slender build, with fairly long legs, straight back, and ribs deep and well sprung, although larger goats than this ideal are acceptable if they are not coarse. The neck should be slender without tassels, but variations from the ideal include neck tassels, long coat and shades of gold in coat colour from pale honey cream to deep gold.

At present the Golden Guernsey has such persistent faults as weak hocks and ugly udders, and improvement is slow because of its small numbers, which make it hard to cull rigorously, but it does offer an opportunity to the serious breeder. Its early ancestors may have been of Swiss or Mediterranean origin.

(8) British

In this section of the B.G.S. *Herd Book* are registered pedigree cross-breeds. The arrangement is almost unique in British stock-breeding, but exceedingly useful; the hybrid poultry breeders and the pig breeders have found the need for a similar arrangement but recently. To the goat breeder it provides a useful pool of genetical characteristics on which to draw for the improvement of existing breeds and the development of new ones, and it gives guidance in cross-breeding for maximum production. The section contains many of the heaviest milkers in the country.

The cow dairy farmer who decides to stock his farm with, say, Jerseys, and obtains his foundation stock from a number of different breeders, will find himself with a reasonably uniform herd. But the goat dairyman who stocks with, say, British Saanens, obtained from various sources, is likely to accumulate a herd of goats of all shapes and sizes and characteristics, though almost all of them would be white in colour. While the breeds do have distinctive utility characteristics, to obtain a reasonably uniform herd it is necessary to select

Table 6. Relative popularity of breeds in Britain, 1945–78 (Numbers and percentages of registrations in the B.G.S. *Herd Book*)

		1945	1950	1955	1960	1965	1970	1974	1978
Anglo-Nubian	No.	58	118	95	103	148	224	454	870
	%	(1·9)	(4·0)	(7·9)	(9·9)	(11·3)	(12·9)	(14·7)	(10·9)
British Saanen	No.	683	587	233	180	213	329	470	1,187
	%	(22·4)	(20·1)	(19·3)	(17·3)	(16·3)	(19·0)	(15·3)	(14·9)
Saanen	No.	132	108	70	58	74	70	94	156
	%	(4·3)	(3·7)	(5·8)	(5·6)	(5·7)	(4·0)	(3·1)	(1·9)
British Toggenburg	No.	154	169	89	148	187	246	343	829
	%	(5·1)	(5·8)	(7·4)	(14·2)	(14·3)	(14·2)	(11·1)	(10·5)
Toggenburg	No.	29	30	21	24	49	32	46	101
	%	(1·0)	(1·0)	(1·7)	(2·3)	(3·8)*	(1·8)	(1·5)	(1·3)
British Alpine	No.	376	316	123	99	115	128	197	356
	%	(12·3)	(10·8)	(10·2)	(9·5)	(8·8)	(7·4)	(6·4)	(4·5)
British	No.	640	737	334	274	308	393	686	1,519
	%	(21·0)	(25·3)	(27·7)	(26·3)	(23·6)	(22·7)	(22·3)	(19·2)
Foundation Book	No.	365	310	107	52	85	92	262	854
	%	(12·0)	(10·6)	(8·9)	(5·0)	(6·5)	(5·3)	(8·5)	(10·8)
Supplementary Register	No.	610	541	133	102	125	219	504	1,661
	%	(20·0)	(18·6)	(11·0)	(9·8)	(9·6)	(12·6)	(16·4)	(20·9)
Identification Register	No.	–	–	–	–	–	–	–	325
	%	–	–	–	–	–	–	–	(4·1)
Golden Guernsey (G)	No.	–	–	–	–	–	–	23	–
	%	–	–	–	–	–	–	(0·7)	–
S.R. (G)	No.	–	–	–	–	–	–	–	1
	%	–	–	–	–	–	–	–	(0·12)
Total		3,047	2,916	1,205	1,040	1,304	1,733	3,079	7,924

* above average.

from within the chosen breed the strains and individuals of the desired type.

Some of the utility characteristics—udder shape, long legs, coat quality, degree of fatness—declare themselves openly to eye and touch. To discern the outward and visible signs of the hidden qualities that make an efficient milker is the main task of the show judge; and the necessary knowledge cannot be compressed into a paragraph.

But as a coarse pointer: if you want great milk capacity, look for a long lean head, interested eyes, and lively movements, a long, slim neck and gently sloping rump; feel for a fine, pliable skin with a smooth and lustrous coat, for a large, elastic udder and knobbly milk veins; if possible, consult the milk records of the goat or her dam and sire's dam.

Food capacity is largely a matter of sheer size, with particular emphasis on the depth and width and spread of the ribs and the development of a deep wedge-shaped profile; but a nervous, fidgety goat eats more than a placid one. Most goats that sag in the middle cannot carry a great load of fodder over rough grazings—but Anglo-Nubians have their legs more widely set than the other breeds and, like a suspension bridge, they are designed to sag. A long jaw is a help to a big appetite, and long legs to a consumer of branches.

Food-conversion efficiency can be measured only approximately by the eye, being lowest in the short, thick, fat goat and highest in the long, lean, glossy one. Goats carry their basic fat deposit on the chest and as a quilt over the belly. If your fingers find fat on the back and rump of a milker, the milk pail is being robbed to clothe the goat—but on goatlings and kids, back fat is a reserve for their first lactation. In assessing the goat's tendency to fat it should always be remembered that a dry diet with a restricted or unappetizing water supply will fatten any goat, while the nutrients that accompany a big fluid intake always tend towards the milk pail.

The detection of these basic utility qualities in the genetic make-up of the male goat is not a suitable task for the eye and hand of inspection—not even if the eye and hand be those of an expert. In selecting a herd sire, his progeny are the only reliable guide to his quality; his pedigree may give grounds for reasonable hope.

While his skin quality, character and conformation have a similar significance to the parallel qualities in the female, their appearance in the male goat may be easily confounded by secondary sexual characteristics and accidents of rearing.

The operation of the scent glands on the he-goat's skin, and his habit of splaying his front legs and neck, occasionally set up in the housed billy a kind of dermatitis which converts a naturally soft and pliable skin into an unpleasant hairless corrugated leather. The practice of feeding male kids large quantities of milk until they are far past the natural weaning age is liable to result in a distorted growth—great length and bone and great depth due to the sag of the milk-filled stomachs, but far less width and spread of rib than that to which his inheritance entitles him. The need under common circumstances to keep the males in a small enclosure and carry their fodder to them may lead to them receiving a rather concentrated ration, which accentuates the appearance of narrowness without in any way affecting what inherited capacity they may have for breeding wide-ribbed kids.

There are also certain characteristics that may be required of the herd sire, but are not particularly wanted in milkers. Exceptional docility, for example, is in common demand. A well exercised he-goat accustomed to kindly but firm control, is not a dangerous beast. But if chained and penned for life and handled with some fear and disgust, any but the poorest in spirit becomes unmanageable. Though a very soft-tempered male seldom breeds outstanding milkers, or a temperament suited to free-range life, poor-spirited and effeminate billies are justifiably popular with those who lack the facilities for keeping a more vigorous beast under control. The Saanen breed generally produces the quietest males, and the fleshier strains of British Saanen are usually temperate in habit.

The odour of billy is universally unpopular, a menace to the production of palatable milk, and an all-round nuisance. Its seasonal production is a secondary sexual characteristic of the goats of cool regions, in which there is a limited breeding season. In tropical countries, where the male remains sexually potent all year round, scentless males are not uncommon and the odour of the odoriferous is milder. The tropical ancestry of the Anglo-Nubian has resulted in some strains of the breed producing very mildly scented males and there seems to be no theoretical reason why this highly desirable characteristic should not be further propagated. Effeminate males of other breeds are often mildly scented too; but in the Swiss breeds, such males frequently prove unsatisfactory breeders.

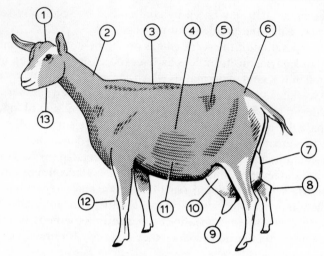

Fig. 15. Points of the productive goat

(1) Long, lean head with lively expression.
(2) Long, lean, silky-skinned neck.
(3) Strong, straight, muscular back.
(4) Deep, wide-sprung ribs, last rib curving back.
(5) Hollow in front of hips bespeaks capacity of digestive organs.
(6) Long, gently sloping rump to support heavy vessel.
(7) Capacious vessel, broadly based on the body and with fine, elastic skin, swells the rear profile.
(8) Hocks are sufficiently straight to avoid bruising vessel when the goat walks.
(9) Teats are hand-sized and distinct from the rest of the vessel; the thicker they are, the quicker she will be to milk.
(10) Extension of vessel forward is the goat's safeguard against udder chills.
(11) Under the belly you can feel, if not always see, the big, knobbly, milk veins.
(12) Strong, clean bone in front legs.
(13) Long, powerful jaw to match a big appetite.

The smell of billy is a real problem of practical importance which tends to get overlooked because it is also a joke. Contamination of the clothing of the billy's attendant is unavoidable, and the regular attendant becomes inured and insensitive to the smell after some months. But whenever contaminated clothing comes within the range of the public's nostrils or an open can of milk, damage is done to the goats'-milk industry. It is an elementary precaution to handle the billy and milk the goats in distinctively different overalls; clothing can be fairly effectively decontaminated by hanging it in acrid smoke, such as is produced by smouldering oak sawdust, or by

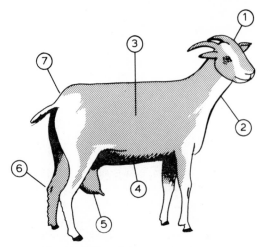

Fig. 16. Points of the unproductive goat

(1) Short, convex face, and more or less upturned nose.
(2) Short, coarse neck.
(3) Shallow, straight ribs, and brick-shaped profile.
(4) Small belly.
(5) Small, tough-skinned vessel, purse-shaped with little, finger-breadth teats.
(6) Hocks nearly together as she walks.
(7) Short, steep rump.

She is dear as a present to 'a good home', unless she goes straight into the pot.

the bee-smoker's corrugated cardboard. A radical solution of the problem is offered at the end of Chapter 9 (p. 222).

In many free-range herds, the male runs out with the rest of the flock, at least during the summer. Special considerations arise in selecting a male for this routine. Under such natural conditions the social structure of the wild goat flock is resurrected, and the male leads and controls the flock's grazing if he is able to, especially during spring and early summer. In the wild herd the he-goat's size and his ever-growing hair and horns call for a supply of nutrients and minerals which differs little from the requirements of the milkers. The short-haired, hornless size of the domesticated herd has needs insignificant in comparison with those of high-yielding milkers, and his leadership of the foraging expedition is sheer misguidance. Where the flock and range is small, this misguidance results in lowered condition and yield; where the flock is large and the range extensive, it results in a splitting of the herd on range, the milkers generally following a leading she-goat and the dry stock following the billy; this division of the flock multiplies the problems of control and management. But the problem is minimized by insisting that the male who joins the flock on free range is always a young one. The he-goat goes on growing until he is two and a half years old; so long as he is fleshing a large and growing frame, his dietetic needs are not so far below those of the milkers; until he is old enough to throw his weight around effectively, his efforts to misguide the flock will be thwarted by the leading milker.

Under few circumstances are horns desirable, either in the male or in his progeny. Only when goats are closely confined or tethered in places liable to regular invasion by small boys and uncontrolled dogs may the balance of advantage lie with the horned goat. Not that horns are dangerous to the goats' attendant; indeed, they provide useful handles. But they are very dangerous indeed to the udders of other goats, and a tragedy is ultimately inevitable where horned goats are run together, and more speedily inevitable where horned and hornless goats associate. The use of a disbudded or de-horned male is often advisable, but only if every horned kid is disbudded.

By far the most satisfactory male for any goat dairy farmer (as opposed to pedigree breeder) to select is the one standing at stud within a reasonable distance of the farm. The dairy farmer is justified in turning a blind eye to considerable defects in any billy he can use but is not called upon to keep.

Artificial insemination of goats has been practised in the Soviet Union, in India, Denmark and France. In Denmark, the goat breeders inseminated their own goats with semen expertly collected. This scheme achieved up to 80 per cent of successful inseminations, the best results for any A.I. service for goats; but it was destroyed by the Danish Government withdrawing their very modest financial support as a national economy measure.

In France, a state-sponsored insemination service is available in the main goat-breeding areas. Semen is collected as for cattle, diluted to one-quarter of its original concentration, chilled to 3° C, and stored for up to 24 hours. Insemination is carried out by trained staff, using the same techniques as for cattle. Sixty per cent of inseminations are successful; as far as possible, goats are inseminated 12 hours after the first appearance of 'season', this stage having proved the most favourable. The French find a 35 per cent variation in the fertility rate between males, even though the sperm count and motility be comparable. This rather mysterious male fertility factor, and the difficulty in maintaining the potency of goat semen over long storage periods, seem to be the main technical difficulties. As such they are no impediment to a practicable and economic A.I. service for goats.

As a means of stock improvement, A.I. is not of immediate importance in Britain. Improvement of productivity by genetic means proceeds at the rate of about 1 per cent per generation, under good genetic management. Improved kid-rearing and goat-feeding methods could probably raise productivity 20 per cent in one generation.

Chapter Nine

Breeding Problems

The demands that mankind makes of the goat have not required any fundamental alteration of its breeding mechanism, nor distortion of its shape and proportions. In this respect the goat is a great deal more fortunate than its comrades of the farmyard, and consequently remains relatively free from breeding troubles.

The naturally early maturity and high fecundity of the goat is adequate to meet any normal demand for the maintenance or increase of stock. The problem of winter milk supply is solved by allowing the goat to go wild or 'run through' alternate years, and run a twenty-two-month lactation as a matter of course. Even on this routine, an average goat in a cool climate will produce five daughters in eight years—if a dairy cow produces five daughters in a lifetime she is doing well. Consequently, out-of-season breeding with goats has nothing like the utility value of the relative freedom from breeding troubles they enjoy, by retaining their natural sexual cycle and glandular rhythm.

The mechanism which controls the breeding cycle of cows is understood to a limited extent. It is possible, with the hand in the cow's rectum, to feel the whole of the genital organs and the changes that take place in them. The mechanism controlling the breeding cycle in goats is believed to be similar; but the belief cannot be so tangibly confirmed.

The sexual cycle is started by the anterior lobe of the pituitary gland (the 'master gland' below the brain) secreting F.S.H. (follicle-stimulating hormone), a substance that excites the ovaries, at each tip of the horns of the womb, to develop a blister, inside which one of the store of eggs in the ovary rapidly develops. This blister itself secretes oestrogen, a substance that produces the symptoms of oestrus (heat). The womb contracts, the cervix at the mouth of the womb relaxes and opens, the vagina is tensed and lubricated by the discharge of slime. The goat becomes restless, bleating and wagging its tail, with a red and swollen vulva often showing some of the discharge. Vagina and womb are prepared for the entrance of the male sperms, and the goat will 'stand' to the billy.

When the blister in the ovaries reaches its full size, the pituitary produces L.H. (luteinizing hormone), causing the blister to burst and the mature egg to start on its passage down the long, twisted fallopian tubes which join the ovaries to each horn of the womb. The broken walls of the blister then grow lutein, a substance which forms a kind of temporary gland that secretes progesterone, which has the opposite effect to that of oestrogen. The outward symptoms of 'heat' subside, the vagina relaxes and dries, the cervix closes to seal the womb, and the womb relaxes and is richly supplied with blood. If service has been effected, the egg on its passage down the fallopian tubes encounters a sperm and is fertilized; arriving in the womb it finds the place prepared for it by the action of progesterone, settles down and develops. The lutein remains, continuing to secrete progesterone until the foetus is mature. Then the lutein is reabsorbed, and, at the stimulus of the pituitary, the ovaries once more secrete oestrogen, which relaxes the cervix, lubricates the vagina and contracts the womb to expel the kid, at the same time inducing the restlessness, and often the other symptoms of heat, in the goat's behaviour. When the kid is born, the secretion of oestrogen ceases.

If the egg is not fertilized in its passage down the fallopian tubes, the lutein persists, secreting progesterone, for about ten days; then it shrinks away and F.S.H. is again secreted by the pituitary to start the cycle over again.

What stimulates the pituitary to start the whole process is obscure. But the main function of the pituitary in this matter is to ensure that kids are born only if and when conditions are suitable for their survival. Under tropical conditions, where seasonal changes of climate and vegetation are immaterial, the goat's breeding season extends over the twelve months of the year; lactations are short where Nature provides abundant food for early weaning, and two or more pregnancies a year are normal. Normal health is the only stimulus needed to maintain the cycle. Farther from the

Equator, seasons become more pronounced. Generally speaking, in the Northern Hemisphere, to which the goat is native, the spring months of April and May offer the kid the most auspicious welcome. As the goat, like the sheep, carries its young for five months, the months of October and November are ideal for starting the breeding season. Various factors signify the arrival of these months: diminishing length of day, falling temperatures, herbage growing drier and tougher. Each of these factors plays its part in stimulating the onset of the breeding season. In Kenya, where the length of day does not change greatly, but there are two periods of rain each year, followed by lush growth and drought, the hardening of the herbage with the drought starts the breeding season, and the kids are born after the rains; so there are two breeding seasons a year. Farther north, and throughout Europe, the dominant factor in the onset of the breeding season is the diminishing length of day.

Our goats are derived from a wild stock native to the Mediterranean basin, and their glandular control is still designed to produce kids in April and May in these latitudes. The onset of the breeding season in southern France and the Balkans is in mid-October, and the kids arrive in April. But the critical rate of change in the length of the day takes place closer to midsummer the nearer we go towards the North Pole. In the extreme north of Scotland the length of the day is diminishing as rapidly at the end of July as it is in mid-October in Marseilles. Consequently, the goat's breeding season in the north of Scotland starts at a time of year which results in the production of kids in the first days of January, and a very cold welcome they get.

However, the climatic variations are overcome by modern science, which, apparently without subsequent ill effects, brings a goat in season at any time of the year, a vital factor for the commercial herd owner, who cannot operate effectively unless kiddings are staggered to keep up a continuous milk supply. Likewise, a kid, if accidentally mated, can be effectively treated by safe veterinary methods. At one time the risk of cystic ovary development, whereby she developed male tendencies and became a non-breeder, presented a greater risk than that attendant on the far-reaching effects of parturition for an immature animal.

For three weeks after mating, the fertilized egg lies free in the uterus, nourished by a secretion of the uterine glands. After three weeks the outer, trophoblast, layer of the developing egg eats into the mother's tissues and forms an attachment—the placenta. For the next ten weeks the trophoblast cells act like a sort of cancer, extracting all the needs of the growing embryo from the mother's tissues, whether she can spare them or not. Level of feeding of the mother during this period has no effect whatever on the growth of the embryo, which extends its attachment to cover the area of uterus wall available. Then eight weeks before kidding, the trophoblast cells die off, and the foetus is fed by transfusion from its mother's blood-stream into its own. The quantity of transfusion depends on the area of the placental attachment, which will be less than the optimum when the goat is too small, or too fat, or carrying too many kids for her size. The quality of the transfusion the foetus receives will depend on the current diet of the mother and her metabolizable body reserves.

The end product of the sexual cycle is not only the kid, but lactation. A number of different hormones control the development

Table 7: Goat Gestation

| *Mated in* | *Will kid on* | |
	(Mating date less no. below)	
August	January	−3
September	February	−3
October	March	−1 or −2
November	April	−1 or −2
December	May	−1 or −2
January	June	−1 or −2
February	July	0 or −1
March	August	−3
April	September	−3
May	October	−3
June	November	−3
July	December	−3

This table is based on an average gestation period of 150 days, and the date a goat may be expected to kid is calculated by subtracting from the mating date the number indicated in the table. A goat mated on 10 September, for instance, would be expected to kid about 7 February, whereas a goat mated on 17 April should kid about 14 September. Where February falls during the 150 days, allowance is made for leap years.

of the udder and the secretion of milk; both oestrogen and progesterone play a part in normal development of the udder, and the secretion of the thyroid gland, which is poured out in increased quantities during pregnancy and spring, has the effect of diverting nutrients towards the womb and udder, thus assisting milk production. Oestrogen itself inhibits the secretion of milk, so most goats show a slight fall in yield when they come in 'heat'. But a falling level of oestrogen in the blood, such as occurs at the end of pregnancy, triggers the chemical mechanism which starts the goat milking. In barren goats, too, the level of oestrogen in the blood falls at the end of the breeding season in early spring, when the thyroid is most active; consequently they often become 'maiden' milkers.

The male sex organs are a little more complicated than outward appearance suggests. The testicles in which the sperms are produced consist of a number of minute tubes, converging into a central tube. The sperms are formed in the small tubes, move into the central tube and thence into the epididymis, a very long convoluted tube in which the sperms mature. From the epididymis they emerge into the cord connecting the epididymis with the seminal vesicle above the bladder, where the sperms are diluted in the seminal fluid which activates them—the seminal fluid being produced by a number of contributory glands.

The pituitary of the male goat, like that of the female, produces both F.S.H. and L.H. L.H. stimulates the testicles to produce both sperms and the hormone *testosterone,* which gives the billy his typically male appearance and behaviour. A deficient secretion of L.H. is symptomized by a feminine appearance, sluggish service and low fertility, and is a condition which may be inherited or result from faulty feeding.

Although the fertility of goats is occasionally impaired by errors of diet and management, or by disease, the main fertility troubles are connected with the problem of hermaphroditism. Of all the aspects of goat husbandry, this problem of hermaphroditism seems to be quite the most interesting to scientists. During recent years, geneticists, pathologists, and microbiologists from at least five different countries have published papers on the subject. Their theories remain inconclusive; but all their facts support the same practical breeding policy, and are consistent with the ecological explanation of the subject offered in the first edition of *Goat Husbandry* in 1957.

A hornless or 'polled' mutant turns up from time to time in all the horned species—cattle, deer, sheep, goats, etc. It is a tentative measure of disarmament, a saving of expense on defensive weapons; provided the defensive weapons have ceased to serve a purpose, the mutation is likely to prove successful. But in animal societies such as those of the red deer and the mountain goat, in which the males select the pastures for the herd, and for whom the main natural enemy is starvation, though female horns be irrelevant, the annual renewal of the stag's antlers and the ever-growing burden of the male goat's horns are society's sole guarantee that the male's choice of pasture matches the needs of the mothers (see p. 198). Although the stag does not lead his family party as the male goat does, it is the summer parties of antler-growing stags that pioneer new territory for the expanding red-deer herd. For both deer and goats, a hornless male may be social disaster. Inevitably the hornless male will keep in better condition, and be more mobile, than his horned or antlered rival. During the breeding season his advantage must ensure him a disproportionate share of services; if fully fertile, he must reproduce his kind, to the destruction of the flock. The polled stag or 'hummel' is always to be found in the deer forest; he is to the fore at the 'rut', and more than a match for the antlered stags; yet even in the most neglected forests the percentage of hummels does not greatly increase; and on the best-managed estates, where hummels are rigorously weeded out, the proportion of hummels remains much the same.

In the hill country, one year in seven is a year of disaster, when the snow lies too deep and winter lasts too long. When the new grass starts to grow at last, it grows greenest around the corpses of the antlered stags. The hummel survives, wintering on his fat; in the autumn of that year the survival of the herd may largely depend upon such fertility as the hummel has. Mountain pastures are a delicately balanced community of plant and animal life: alter one element and the whole kaleidoscope changes; remove half the deer, and the vegetation coarsens; let the vegetation grow rank, and the goat moves in. A species must make a rapid recovery from disaster to retain a footing in its habitat. The hummel's fertility needs to be exceptional.

The only sort of hornless mutation which is going to succeed among mountain goats or mountain deer must satisfy two requirements: (1) under normal conditions and natural selection, the proportion or hornless males must not exceed about 5 per cent; (2) increased use of hornless males should boost herd fertility.

The hornless mutation that has survived among domesticated goats in Europe almost certainly originated among the mountain goats of Switzerland; indeed all the investigations of the factor, in Israel, Britain, Germany and Japan, have been carried out on descendants of exported Swiss Saanens. This mutation meets the ecological requirements of the mountain-goat flock in the following way:

The factor for hornlessness is inherited as a Mendelian dominant: if the gene is present, there are no horns, if absent, there are normal

The horned goat is *homozygous*—that is, the factor for horns is inherited from both parents. The factor for horns, being recessive, would be apparently suppressed if a factor for hornlessness were inherited from either parent. Linked to the factor for horns is a dominant factor for normal sexual development. The horned goat passes on to all its offspring a 'recessive' factor for horns and a 'dominant' factor for normal sexual development.

The homozygous hornless goat inherits from both parents a factor for hornlessness which is dominant, linked with a factor which tends to change the foetus in the womb into a male. If the foetus started male it is unaffected; if it started female it will be born apparently female, or obviously intersex—but always sterile. The sex-change factor is recessive, effective only in the homozygous state. There are no homozygous hornless females capable of breeding, but the males pass on the dominant hornlessness and recessive sex-change factor to all offspring.

The heterozygous hornless goat inherits from one parent the recessive factor for horns, linked with the dominant factor for normal sexual development, and from the other parent the dominant factor for hornlessness with the recessive factor for foetal sex-change. The goat appears hornless, and its sexual development is normal; but it can pass to its offspring a factor either for horns or hornlessness, each with its linked factor for normal or abnormal sexual development. All hornless females capable of breeding are heterozygous.

Fig. 17. Hornlessness and hermaphroditism (a) The three genotypes

Fig. 17. Hornlessness and hermaphroditism (b) Safe matings

horns; there is no half-way stage. But hornlessness is closely linked with the factor which modifies fertility. In heterozygous form in females it causes an increase of 5 per cent in fertility by increasing the incidence of twinning and triplets; in homozygous forms in females it causes a partial, or apparently complete, pre-natal sex change; no true homozygous females are born, but a mixture of pseudo-females, pseudo-hermaphrodites, and pseudo-males; all, of course, hornless, and all infertile. Neither in homozygous nor heterozygous form is this factor known to affect the potency of true males, one way or the other. The infertility of the pseudo-males shows itself in two distinct forms: either under-sized testicles, or a blockage in the seminal ducts in both testicles, causing the accumulation of sperm to form a small abcess. Few or no viable sperms are produced in either case.

The practical implications of this state of affairs are perfectly straightforward. There is no possibility of establishing a truly hornless breed of goats, as no true-breeding hornless females are born. (To be exact, possibly one in a thousand hornless females is true-breeding, but she goes unrecognized.)

But it is still worth propagating the hornless factor, for convenience and for economy; and for the slightly greater fecundity, where that is desired. None of the progeny of a horned/hornless mating will be sexually abnormal, and half or more should be hornless. In a hornless/hornless mating, if the male is heterozygous, one in four of the female embryos will become hermaphrodites or pseudo-males; if the male is homozygous, half the female embryos will be affected.

The loss caused by hermaphroditism arises, not so much from the visibly abnormal kid, which is destroyed at birth, but from the pseudo-male or pseudo-female, which is reared to maturity before its worthlessness is discovered. So it is seldom worth rearing the apparent female offspring of a homozygous male and a hornless female; the chances of normality are no more than evens. To exclude homozygous hornless males, it is necessary in practice to exclude all hornless males, both of whose parents were hornless, until they have sired a horned kid from a horned female, and so proved their heterozygous state. Rearing the hornless offspring of the mating of a hornless female with a heterozygous hornless male is still not a commercial proposition: setting the cost of rearing against the selling price of the in-kid goat or buckling, the loss on rearing one hermaphrodite or pseudo-male eliminates the profit on several normal kids.

Fig. 17 summarizes the situation in conventional symbols.

The most important dietic causes of infertility in male goats are 'high-calcium' phosphorus deficiency, vitamin-A deficiency, iodine deficiency and copper deficiency, to which falls to be added the somewhat mysterious effects of eating mangolds, which, when in season, are harmless to the female.

Vitamin-A deficiency and iodine deficiency impair the billy's keenness for service. Phosphorus and copper deficiency do not

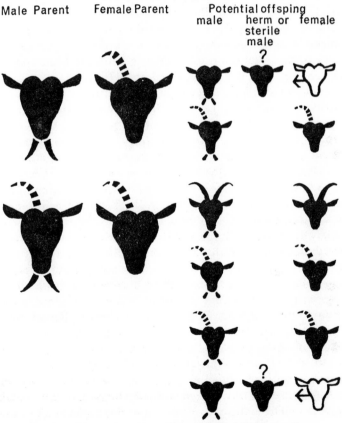

Fig. 17. Hornlessness and hermaphroditism (c) Risky matings

affect his willingness, but lower the quality of the semen. The cause
and treatment of these deficiencies are dealt with in the chapter on
Feeding. The danger of high-calcium phosphorus deficiency is, of
course, enhanced if the billy is fed a dry diet and supplied with large
quantities of hard (calcium-rich) drinking water.

If the herd male is willing to serve, and his semen when tested
shows a normal content of active sperms, the cause of herd infertility must be sought among the females. In this search the goat farmer
has less ground to cover than the cattleman. Goats are effectively
free from the highly infectious diseases that afflict the genital tract
of cattle. Occasional infection of goats with contagious abortion
occurs where the disease is prevalent among associated cattle; but
goats have a high resistance to the disease, and the worst that can be
expected of it is an occasional abortion late in pregnancy.

Where the cause of a herd-breeding problem lies with the
females, it almost always lies with the feeding. Vitamin-A deficiency is only likely to occur following herd infestations by liver
fluke and coccidia, the symptoms of which (p. 236) are unlikely to
pass unobserved. This deficiency does not affect the regularity of
oestrus; but causes changes in the lining of the uterus which make
conception difficult, and early abortion probable. Copper deficiency is confined to notorious localities, and is symptomized by
irregular periods between 'heats', and return to service at irregular
intervals. Iodine deficiency is quite common (see Chapter 6,
p. 115). As a breeding problem it shows itself in failure to come in
heat, or irregular periods between heats; if conception is achieved
the kids are usually carried full term, but most of the females and
some of the males are born weak or dead; the vessels of first kidders
show disappointingly little development. Where failure or irregular-
ity in coming in heat is the problem, iodine deficiency is the most
likely cause; if kale is being fed to a high-yielding herd with this
problem, iodine deficiency is even more likely to be the cause.

Phosphorus deficiency is also common in high-yielding herds. As
a breeding problem in a fair-sized herd it is not hard to spot, because
while some of the goats will be on heat almost perpetually, and even
start to develop male characteristics, other members of the herd will
refuse to come in heat at all, or at best but very seldom. It is less easy
to spot among two or three goats, all of whom may show the same
symptoms. The deficiency causes the development of a cystic
ovary—that is, the follicle in the ovary turns into a cyst which

persists in secreting either oestrogen (in which case the goat is perpetually in heat) or progesterone (in which case she comes in heat only once, or not at all). In either case, service is ineffectual.

Unfortunately, a tendency to cystic ovaries can be an inherited factor; so the appearance of the typical symptoms in a closely related herd may be due to either phosphorus deficiency or inherent fault.

To distinguish between these two possible causes, note that the form of phosphorus deficiency that afflicts high-yielding herds is due, not to a poverty of phosphorus in the diet, but to a superabundance of lime, which prevents the digestion of available phosphorus. The typical diet which may produce it is one consisting of legume hay as the basic roughage, accompanied by a large percentage of sugar-beet pulp among the concentrates, and supplemented by an ad lib supply of seaweed meal. If such diet is general to the whole herd, then it is liable to produce not only adult breeding problems, but also a kind of chronic cramp in young stock, especially in very early spring. The picture of limping goatlings on the one hand, and miscellaneous breeding problems on the other, is a conclusive pointer to phosphorus deficiency as the cause of both troubles.

While further incidence of phosphorus deficiency can be prevented by change of diet, it will take at least a month of changed diet to remove the symptoms of the deficiency, and a cystic ovary that is well established may be incurable. Veterinary surgeons can treat the condition with some slight hope of success and at very considerable expense, by injections of hormone. Perhaps the most economical and hopeful treatment is that suggested for the establishment of an artificial lactation, which should not be applied until the first flush of spring calls forth the natural aid of the thyroid gland to re-establish normal glandular rhythm.*

When breeding troubles are confined to an occasional individual in the herd, while the nutritional factors outlined above must still be borne in mind, the possibility of hereditary defect is very great in any goat that is not a proven breeder. Hermaphrodites range from the almost perfect male to the almost perfect female; though the intermediate types may be fairly easily recognizable, the extremes of the range suffer from imperfections which can only be discovered

* And only to goats which do not come on heat regularly.

at post-mortem examination. The proven breeder who fails to hold
to service is more likely to be a victim of nutritional deficiencies if
she is a heavy milker, or to be suffering from cystic ovary or over-fat
condition if she is a poor producer.

Fault-finding Chart

This chart lists solutions in the order in which they should be tried.

(1) HEAT PERIODS NORMAL—GOATS RETURN TO SERVICE.
DOES THE MALE SERVE?

If he does:
Have his semen tested

If he does not:
Have a look at his feet
Give more greens, carrots, vitamin A
Give iodized salt, seaweed meal
Replace him

If the sperm count is low:
Give soft drinking water, non-legume hay, no beet pulp or seaweed meal; but oats and bran and high-phosphate supplement
Give copper supplement
Replace him

If the sperm count is normal:
Give greens, carrots, vitamin A
Can he be infested with liver fluke? Or coccidia? If so, treat

Replace him

(2) HEAT TOO FREQUENT OR PROTRACTED—GOATS RETURN
TO SERVICE
Give soft drinking water, non-legume hay, no beet pulps, no seaweed meal; but oats and bran and high-phosphate supplement
If there is no improvement after one month, goats possibly have cystic ovaries: consult vet for expensive treatment

(3) HEAT IRREGULAR, INFREQUENT OR ABSENT—GOATS
RETURN TO SERVICE
Give housed goats range with a billy, or yard them together
Give soft drinking water, non-legume hay, etc.
Give iodized mineral supplement, including copper
Suspect cystic ovary: consult vet, or try inducing artificial lactation in late April

(4) Goats Abort in Late Pregnancy
Suspect and avoid poisoning, especially with vermifuges, including garlic
Suspect rough management and narrow doorways
Decrease feeding during early pregnancy, and increase during last eight weeks
Avoid either over- or underfeeding

(5) Goats Abort in Early Pregnancy
Are they infested with liver fluke?
Or with coccidia? If so, treat
Or have they been so infested during previous summer? If so, give more greens, carrots, vitamin-A supplement

(6) Difficult Kiddings
The goat strains vigorously and the kids are strong:
Decrease feeding in early pregnancy; do not give cod-liver oil
The goat does not strain vigorously; kids are weak or dead:
Increase feeding during last eight weeks of pregnancy
Give iodized mineral supplement or seaweed meal

(7) Goat 'Turns', i.e. Comes in Season, After Being Served by Fertile Male
Take an enamel jug and put 1 tablespoon of bicarbonate of soda—add rather more than 1 pint (0·6 litres) of boiled water, cool to blood heat. Douche the goat with an enema with vaginal attachment. Get the water running (to expel air bubbles) before inserting, and have an assistant to hold the goat, then press down on the goat's back as she may expel the water before it reaches its objective. Twenty minutes later have her served. This corrects acidity and she may 'hold' or 'settle', i.e. conceive. The treatment has been successful not only with goats but with other stock, particularly mares.

(8) Goat 'Turns' after Mating to Unproved Male Kid
Get the kid's semen tested for active sperm. In order to test he must serve a goat. If there are no active sperm or only a few, destroy the kid; expensive treatment is useless. It is usually understood that should a male kid prove infertile his purchase price is refunded by the seller.

(9) 'CLOUD BURST' OR FALSE PREGNANCY

Occasionally a goat, whether served or not, appears to be pregnant, but eventually 'gives birth' to a large quantity of water and nothing else, and bleats for her non-existent kid. The cause is unknown; the goat breeds normally after the discharge has ceased.

KIDDING

Perhaps the greatest source of anxiety concerning the goat in kid is uncertainty of the day on which the kid will be born. The average period of gestation, from date of service to date of kidding is 150 days. The normal range is from 143 to 157 days, which gives the herdsman who has noted service dates plenty of room for anxiety; if service dates have not been noted, the goat will keep her attendant guessing for a month, and spring a surprise in the end. The only herdsman who is never taken by surprise is the one who passes his hand inquisitively over the goat's right side every morning and evening, as a matter of routine.

In the pregnant goat, as she approaches her term, the kid can be seen and felt to move in the bulge on her right side. So long as that is the case, there is little likelihood of kidding within twelve hours.

The process of birth is started by the goat's ovaries secreting oestrogen; under the influence of this hormone, the muscular walls of the womb, which have remained flabby and relaxed throughout pregnancy, become hard and tense. The kid or kids which have kicked about in a flabby bag suddenly find themselves in a strait-jacket. If they do not lie still, their movements are so little felt by an outsider that they seem to do so. If the goats have been handled regularly there is no possible room for doubt about this change; if the goats have not been handled regularly the change will not be discovered by a panic-stricken last-minute exploration.

Any time from eight to twenty-four hours after the womb has tensed, in a normal birth, one of the kids is forced up into the neck of the womb. This movement causes the bulge in the right side to subside, and tilts up the sloping plates of the rump to a more horizontal angle. The change in appearance can usually be noticed if one is looking for it; but if the kids are many or big, the subsidence in the right side is not very obvious. Once this stage has been reached, the first kid may be born within two or three hours.

Other signs of imminent kidding are a widening of the space between the pin-bones; the appearance of deep hollows on either side of the tail, and rapid filling of the udder. But in old goats, and in goats carrying a heavy cargo of kids, the space between the pin-bones may ring a false alarm weeks before kidding; in heavy milkers, the udder is often quite tight and shiny ten days before kidding, and pre-milking may be necessary.

The secretion of oestrogen which tenses the walls of the womb also affects the behaviour of the goat. A first kidder will often behave very much as if she were in season; an older goat is more likely to appear merely fussy and restless in the shed. But let the goats out of the shed together, and the goat which is about to kid is no longer a member of the herd, no longer follows her leader, but maintains a purposeful and divergent course quite foreign to her normal behaviour. The kidding goat in a free-range herd always gives fair warning, if there is an eye to see it.

Like sheep, goats seem able to postpone or accelerate the birth of their kids to take advantage of good weather; older goats use their discretion in the matter more notably than first kidders.

The first good kidding day which comes within a fortnight in which the kids are due is the normal choice of an experienced mother. A good kidding day is a mild and humid one, with a minimum of wind; lambs and kids are best born in a Scotch mist, where loss of heat by evaporation of the natal slime is at a minimum.

If the day is good and the terrain at all suitable, the best kidding pen is the secluded hollow that the goat will choose for herself, which is more hygienic than anything under a roof, and freer of hazards than anything with a flat floor and four walls. Most goats will choose a place under the cover of trees or rock; an indiscreet first kidder may fail to do so, and should be encouraged to think again; shelter from crows, showers and gulls is desirable. Otherwise human assistance is best limited to watching from a distance. Company of any kind, even of the dearest, is not desired, and it will be many hours before it is needed out of doors, even if things go wrong.

It is not everyone who has free range for goats, or would welcome the idea of supervising kiddings outside. In fact, a goat may refuse to go out with the herd if she is ready to kid that morning. She should have a clean box, with a good covering of short straw which will not get entangled with her restless movements. Long straw is awkward; and a layer of sawdust under short straw, while it absorbs moisture, needs to be completely covered.

When kidding has started, there is a discharge, at first colourless. When birth is fairly imminent, this changes, becomes thick and white, and is a sure indication kidding has actually begun. The goat paws her bedding; lies down with a sigh or grunt, and perhaps strains very slightly; gets up; walks about restlessly; lies down again and strains. She may look round and 'talk' to the kid, which she can probably at this stage smell—this 'mother talk' is something only heard during kidding and for a few hours afterwards; and once heard, it can never be mistaken. These are all normal signs, and you may perhaps go away and, returning half an hour later, hear the sound which tells you that the first kid is now being born. Some goats are extremely vocal during the final stages, when the serious business of hard straining develops; and the shrieks can be unnerving when heard for the first time. First kidders generally make a lot of noise; older goats probably only give a few anguished grunts at the moment of birth. The goat usually lies down, and it is as well to see that her back end is not pressed against the side of the box. She will probably lie stretched out flat, and push with hind legs against the floor or wall to assist the birth. Some goats drop their kids standing up.

Now, if all goes well, the water bag should appear and burst to reveal a foreleg, followed by another with the top of the nose resting on them (see Fig. 18 for this, normal, and other, abnormal, presentations).

The goat may rest for a moment, then, finally, with two huge heaves, expel the rest of the kid. If there is a hold-up, grasp the kid's legs and, as the goat strains, pull them downwards towards her hocks. The first kidder may not realize that she has given birth or recognize the wet, slimy object as her child. Clean the mucus from the kid's nose and mouth and gently slide it round under the mother's nose; she will immediately lick and talk to it. A second, third or even fourth kid may follow at varying intervals. Presently the afterbirths, one for each kid, a double one for identical twins, will come away; this may take some time, and no attempt should be made to pull or interfere with the process. Keep count: the goat may perhaps eat part of them when you have gone for a cup of tea and to fix her an oatmeal drink; but each placenta has a double set of cords,

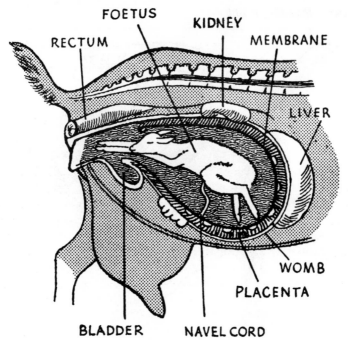

FOETUS
RECTUM
KIDNEY
MEMBRANE
LIVER
WOMB
PLACENTA
BLADDER NAVEL CORD

Fig. 18. Parturition diagrams
(a) Normal presentation.

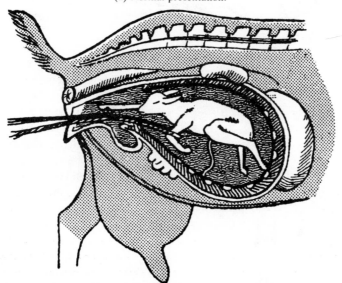

(b) Bring that leg forward—with your finger if possible; the rope shows direction of pull.

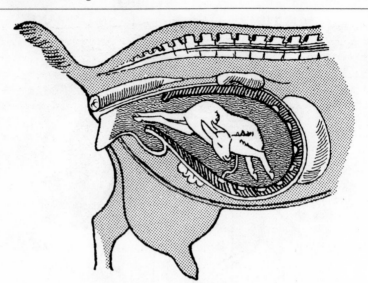

(c) Push the kid right back into the womb, and bring the head forward on to the front legs.

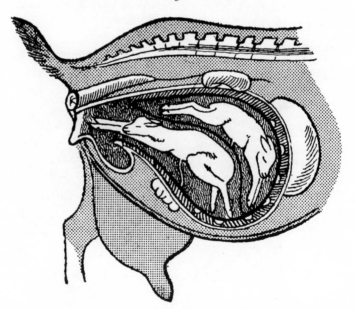

(d) All correct and normal for a twin presentation. But sometimes the hind legs of the second kid get in front of the head of the first. Then you find four legs and a head in the passage. Push it all back, and bring out in the position shown.

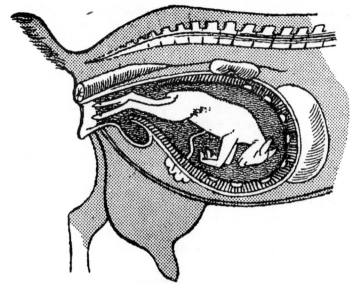

(e) It will come out the way it is; a normal presentation for the second of a pair of twins; but, especially with an elderly goat, ensure that it doesn't linger too long as the shoulders come through.

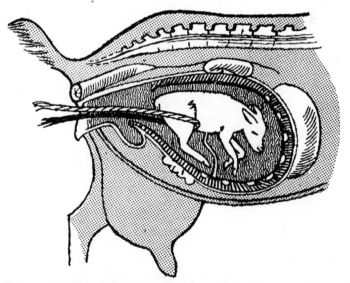

(f) This one doesn't reach the passage at all, but sticks in the mouth of the womb; push him well back into the womb and deliver him in position (e). As before, the rope shows what fingers should do.

and from this evidence you know that everything has come away. A retained afterbirth causes trouble later.

Wash the goat's hind parts with warm water and dry them; remove some of the stained bedding, and top up with fresh. Offer her a warm oatmeal drink made by dissolving a handful of oatmeal in cold water to paste consistency, stirring in a kettleful of boiling water, and allowing it to cool a little. Offer it pretty warm, and she will almost surely drink. Some goats like a tablespoon of molasses added.

Meanwhile, examine the kids for abnormalities: a supernumerary (extra) teat, or a double one which appears just unusually thick but has two holes, can be hereditary, and such kids should be destroyed. One of the more obvious forms of hermaphroditism is seen on the vulva of an apparently female kid as a small, round, pea-sized object. There are other bisexual forms, with both male and female organs on the same kid. Obviously the only course is destruction, which can be done later; the goat settles more happily if at least one kid, preferably two, stay with her for three or four days.

Make sure the kids have sucked; if the second kid is some time coming, you can draw off some milk into a warmed bottle and let the kid take its first feed from a rubber teat. In this way, there will be no hesitation if and when it is to be bottle-fed at four days. Finally, offer the goat a bran mash, give her a rack full of the best available hay, and leave the family in peace for two or three hours.

The above refers to a normal kidding. If after twenty or at most thirty, minutes of hard straining, no progress is made, the first kid is probably lying in one of the awkward positions shown in Fig. 18. There is no point in letting the goat exhaust herself and endanger the kids in a futile endeavour to give birth. The novice learns a lot and gains confidence if a vet, experienced goatkeeper or a shepherd delivers these first kids. Next time, given a little common sense and courage, the job will not appear so formidable.

If you are obliged to tackle the situation yourself, get someone to hold the goat if necessary, or you may have to tie her up. Wash your right arm thoroughly to above the elbow, scrubbing nails clean (long nails are impossible; and anyone liable to be delivering kids should have short nails, not cut at the last minute, but smoothly filed down in readiness). After washing, cover your hand and arm in obstetric cream, bunch the fingers together to a cone and, without touching anything on the way, slide gently into the goat. Now feel around, with a finger at first. The goat strains vigorously and you must work in conjunction with the strains. If the trouble is a turned-back foot you may, with a slightly upward movement, be able to bring it forward (Fig. 18b). For an entire leg or head (Fig. 18c) turned back, or other wrong presentation (such as in Fig. 18f), you need room to work; and this means pushing the kid, or kids right back into the womb. Talk reassuringly to the goat, try to visualize the difficulty, and sort it out intelligently. It is much easier once you get the idea, and if you are able to keep calm and visualize what you can feel. Never pull on anything unless you are absolutely certain you can identify it. Sometimes two kids are in the passage with the fore legs of one and hind legs of another the first thing you find (Fig. 18d), don't pull, push it all back. It helps to handle a young kid, or even a dog, beforehand, so that you can recognize fore from hind, and the position you may have to re-adjust, by touch alone.

Sometimes a goat is all ready to kid, but does no more than stand about looking miserable for some hours. She may have a slimy pinkish discharge or none at all; and she never reaches the hard-straining stage. She eats no hay, and is obviously in trouble. This could be a case for Caesarean section if she has not opened up properly; call the vet.

DISBUDDING

Examine the kids for horns within twenty-four hours. The horned kid has a slightly flatter head and two curls covering the pinpoint horn buds, which can be felt and, if the hair is closely clipped, seen as small, white pimples. The hornless (polled) kid has a slightly domed head, and the curls are absent. If horned, make an appointment with a vet or experienced goatkeeper to disbud them at four days old. Male kids have to be done at two days, but would normally be destroyed or destined for the table or deep freeze at a comparatively early age. All too often the job is deferred for a week or more, which is hard on the kids and makes success less certain. Clip the hair extensively and closely round the horn buds on the third day and, if the kids are to be bottle-fed, remove them from the goat that day and get them used to the rubber teat. Preparing the head previously, and letting the kid go straight to the bottle after disbudding, reduce stress to the absolute minimum.

In expert hands by the hot iron method, disbudding can be done without anaesthetic, and the kid will take the bottle immediately on release. Some vets and breeders prefer to anaesthetize kids for disbudding, in which case a general, rather than local, anaesthetic is advised. Many kids are known to have died from shock caused by the latter. The single-handed operator necessarily puts the kid right out; but, provided it is firmly held between the knee of an assistant, so there is no possibility of a struggle allowing the head to move, and disbudding done not later than the fourth day, the job is over so quickly that the kid will suck, even skip, immediately afterwards.

The iron is either electric or is heated in the fire to cherry red. It is placed exactly over the horn bud; no pressure is exerted, and it is kept on for precisely six seconds—no more, no less. The kid will yell anyway, because it hates being firmly held. There is now a completely flattened area with a small bit of horn sticking up round the perimeter. These bits are seared down with the edge of the iron; the area being already cauterized, this rubbing down is painless. Reheat the iron, and repeat on the other side. Apply gentian violet over the whole area; put the kid down; give it the bottle prepared in readiness; and lay it to rest. The head is very sensitive for the first few weeks: if it gets a knock, the kid will cry out and stagger about in pain for a minute or two. Subsequently the head becomes hard and tough as though naturally hornless.

Some breeders, bringing iron and assistant, will visit goatkeepers by request, or do the kids in their own homes. Disbudding is not an operation for the inexperienced, or for anyone unused to handling animals or lacking strong nerves; it is also not a job to be shirked, shelved or postponed. Horned domestic goats, however gentle, are a potential hazard to children, to each other and to those who handle them: the tip of a horn can all too easily blind an eye.

DEODORIZATION

A technique has been developed by the Robert Ford Laboratory, Pascagoula, Mississippi, U.S.A., to remove a goat's musk glands, making male goats inoffensive, and eliminating the 'goaty' flavour which can affect milk as the result of musk-gland activity in she-goats.

The male goat has a well-known habit of labelling his friends, females and furnishings with his personal stink, by rubbing his head on them. The musk glands are situated in a $\frac{3}{8}$- to $\frac{1}{2}$-in (9 to 13 mm) wide band immediately behind, and along the inside edge of, the base of each horn. In a naturally hornless animal, it is situated similarly around the bony boss. In a disbudded animal, the gland is to be found where it would be if the animal were horned (see Fig. 19).

Musk glands are present in both sexes. Being activated by the presence of male hormone in the blood, their activity is seasonal in the male, and unusual in the female. In an adult, when the normal cover of hair and dirt is removed, the glands are seen as an area of thickened and glistening skin; if active, the skin is raised and folded, forming three corrugations on each side; in a kid, the gland is shiny, and darker that the neighbouring skin.

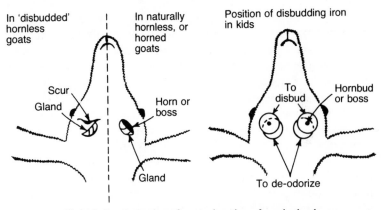

Fig. 19. Deodorization of goats: location of musk glands

The deodorization procedure is a simple extension of the disbudding technique, the glandular area being scorched to a bone-like appearance with a red-hot disbudding iron, in the same way as the horn bud. An adult goat, if hornless, can be deodorized by cauterizing the gland in like manner; if horned, the horns would have to be removed along with the glandular skin, by a long and bloody operation which would be hard to justify. In Britain, all such operations must be carried out under local or general anaesthetic. But in commercial dairy herds and many domestic herds, there would be much advantage for little extra expense if all available kids were deodorized when horned kids were disbudded. The standard disbudding iron can be used, being applied as shown in Fig. 19. Fred Ritson (see p. 361) supply disbudding irons.

There are no serious snags in the procedure. Male fertility is not affected, and kids 'mother' normally. A few animals have small patches of musk gland on other parts of their bodies; these can be located by nose, after shampooing the goat, and cauterized in the same way. Care is needed to avoid scorching through the band of skin that runs between the horns; otherwise, healing is delayed. A goat suspected of producing goaty milk because of an activated musk gland can be tested by rubbing the glandular area with your hand for a few moments and sniffing.

DESTRUCTION OF KIDS

Most male kids and any weaklings, in addition to the abnormal or surplus, will be destroyed at birth or by the fourth day, when the goat has settled down and must be milked dry. Under no circumstances should they be drowned; it is a prolonged and terrible death to which no animal however young, should be subjected. Most goatkeepers use chloroform on a pad of cotton wool; with the kid laid on your knee bring the pad nearer and nearer to its nose as it drops off to sleep. If you breathe too freely on the chloroform yourself, you may start nodding off too! When breathing has completely stopped, lay the kid in the tea chest or outsize carton, renew the chloroform on the pad, put this in a 2-lb (0·9 kg) jam jar and place the jar over the kid's nose. Cover the box and leave till the kid is cold and rigid. Some people just knock the kid over the head—the quickest method if you can do it; but to be sure of death, follow with the jam-jar treatment, or bleed.

BOTTLE FEEDING

Kids are either bottle-fed from the fourth day or left for the dam to rear. She must be milked out morning and evening, or you will not have milk through the winter when the kids are weaned. As goats are bred for long lactation, we should make full use of this ability to milk steadily, with kiddings in alternate years. If the kids are to be bottle-fed, get the goat out on the fourth day with the herd; remove the kids out of sight and sound; clean out the kidding pen; and scatter disinfectant around, especially where the kids lie against the door or walls. Plenty of fresh bedding, first-class hay and, if possible, a rack with green stuff will distract her on return, and she will not fret for the kids. By taking the goat from the kids, never taking the kids from the goat, you have a contented animal settling down to the new lactation; in reverse, you can have a nervous wreck for several days. If the kids cannot be removed to another building, put them in a pen next to the goat so that she can see but not feed them. In a couple of weeks they can run out together without fear that the kids will suck.

Plate 14. First outing

Chapter Ten

Disease and Accident

Why should the writer of a book on animal husbandry be expected to devote a section to disease? The rest of the book is presumably intended to describe the ways and means of keeping the animal in a healthy and productive state. What more is there to say but to repeat what has already been said? If disease arises in spite of the combined efforts of the reader and writer—as it will—then call in the witch doctor, the veterinary surgeon or herbalist, who are technically equipped to deal with the symptoms of disease without removing the conditions that caused it. Their technical equipment is specialized and to be respected; to crib a few of their more popular spells and prescriptions is neither serviceable nor respectful.

If such were the accepted view of the respective roles of stockman and veterinary surgeon, it would be sufficient to supply a few notes on first aid in case of accident and leave the matter at that.

But such is not the generally accepted view. Far from it. The stockman is visualized as defending his charges from a malicious, ever-lurking legion of bacteria, virus and parasite intent on their destruction. The modern stockman is therefore armed with drugs, vaccines, serums and disinfectants to neutralize, suppress and destroy every possible agent of disease. Modern man has made disease after his own image, and determined to kill it.

This is madness, with a monstrous vested interest behind it, and not a scrap of science. The first fact the scientific student learns about bacteria is that they are neutral—they have no desire nor ability to do anything whatsoever, but that which their environment commands. They multiply only if and when their environment is suitable for their multiplication; they remain dormant or die when it ceases to be so. There is no fight, no argument. A virus is as malicious as a lump of salt. Some forms of virus even look like one: the tobacco virus, for instance, can be crystallized, labelled and kept in a glass bottle for as long as you like. Dissolve it, inject a droplet into the right kind of tobacco plant in the right place, and it multiplies, producing the symptoms of disease.

The conditions under which any given strain of bacteria or virus can multiply rapidly are very highly specialized. In the case of disease bacteria, these conditions only exist in one particular type of animal, or in a few allied species. If the symptoms of disease are very severe, or fatal, the conditions which favour the multiplication of the bacteria concerned must be highly abnormal in the animal species affected, or both the species and the disease bacteria would long ago have perished. On the other hand, disease bacteria which produce only mild symptoms may be able to infect under relatively normal conditions.

In fact, any large animal is a focal centre for a vast and diverse population of micro-organisms and 'parasites' whose way of life is intimately bound with that of their host, and which depend on the survival of their host for their own existence. Some of this multitude of micro- and macro- 'guests' are essential to the well-being of their host; others appear to be of little account, one way or the other; some can be helpful or dangerous according to circumstances; others are harmful to the individual they infect, but may act as a useful control of overcrowding when their host is running wild under natural conditions. It seems likely that disease bacteria exist among all these groups.

The relationship between the animal and its 'guests' is well regulated to ensure the survival of both. The more dangerous guests can only gain entry and multiply in the animal body under most unusual circumstances; the less dangerous but potentially harmful guests, by their activity in the animal body, either destroy their own means of sustenance or produce antibodies which control their own multiplication within limits which are safe for the host. The useful members of the community are encouraged to multiply by the circumstances in which they can be useful.

This regulating mechanism works well under normal conditions; if it is to work well for our goats, we must feed, house and manage them to accord as nearly as possible with the life they would lead under favourable natural conditions. In so far as we are obliged to depart from the natural régime, we must expect some breakdown in the mechanism and be prepared to offer additional protection. But the mechanism is generally beneficial: to treat all potentially dangerous bacteria and parasites as enemies is both senseless and expensive.

In handling problems of disease, goatkeepers cannot afford to allow themselves to be bullied. The very large vested interests in the sale of veterinary products take great pains to ensure that the advice that they advertise on the treatment of the main categories of farmstock will not produce adverse results. Occasionally their pains are in vain, but, generally speaking, to follow such advice is unnecessarily expensive rather than disastrous. However, the effect of these drugs and injections upon goats can never be adequately tested, partly because there are too few goats available for full-scale field trials with statistical significance; partly because there are too few goatkeepers to make the research worth while.

On the other hand, crazy and disgusting as the attitude to disease preached by some of these companies may be, many of their products and methods are the best available and perfectly satisfactory if sanely employed.

If we move goats from an area in which contagious pustular dermatitis or louping ill are non-existent into an area where these diseases are prevalent, we are well advised to inject them with vaccine. Such movement over long distances is unnatural to the goat. Goats reared in an infective area will have absorbed antibodies to the disease in their mothers' milk; constant contact with the disease bacteria concerned maintains their resistance. If we introduce goats from a non-infective area we must artificially provide them with antibodies to the disease, so that the natural host–guest relationship may operate. One artificiality demands another—a good rule in the treatment of disease, and a good reason for natural feeding and management.

Similarly, the deep wound from the mislaid hay fork, which gives the deadly tetanus bacteria their rare opportunity of invading the goat, is not one which we should treat with garlic ointment, but with a chemical disinfectant, iodine or hydrogen peroxide. If the district is notorious for tetanus infection, the goat should promptly receive an injection of antitetanus serum.

On the other hand, the bacteria which are associated with most forms of mastitis in goats can only operate when the goat and its udder are already in a diseased condition as a result of bad feeding, bruising, chilling or over-exploitation. The bacteria are helpful scavengers. It is often quite easy to suppress them with drugs; but the ill will out in another way. It is better to treat the cause. But black garget in a goat being sucked by a lamb will be due to the goat being exposed to bacteria to which it has no resistance—one artificiality demands another; bring out the antibiotics.

The general relationship between the goat and her micro-guests is not different in kind from that which exists between the goat and such visible guests as lice, ticks, tape worm, round worm, lung worm and flukes. The goat and her parasites have been associated far too long for the association to be normally dangerous to the goat. However, domestication has done more to disturb the goat's relationship with her parasites than her relationship with her bacterial visitors, so her parasites seem to be considerably more trouble than they are worth; indeed, there is no very obvious reason to suppose that they perform any useful function at all—except the general rule that Nature seldom lets any form of life go on leaving a negative balance behind it indefinitely.

The louse and scurf mite may do some useful scavenging on the unhealthy skin of goats in poor condition, or those who see too little of fresh air and rain—the scurf mite indeed specializes in the unwashed forelegs of an over-housed billy. The liver flukes and coccidia can hardly be expected to make any positive contribution to the goat's well-being, for they have no natural right of access to the goat, who naturally avoids the water-sodden and close-cropped pastures on which these parasites reach their infective stage. But the round worm and lung worm seem to feature eternally in the organs of goats, both wild and domesticated, from the Tropics to the Arctic Circle: their presence in moderate numbers does no harm; it is impossible to say for certain what good it does.

But we may venture a reasonable guess. The picture of the early-spring infestation falling off during summer, and the lesser and briefer autumn infestation of adult goats, will be familiar to all goatkeepers. Worm infestation rises and falls in sympathy with seasonal metabolism—a remarkable state of affairs if the sole function of the parasitic worm were to lower the goat's condition. Infestation rises sharply in early spring, when low temperatures, regular ground frost and the dry condition of the herbage are as hostile as they can be to the development of infective worm larvae—

and again in early autumn, when the free-range goat is beginning to turn her attention to the rougher and drier fodders, and has
no need to eat a morsel of food which grows less than a foot or more
from the ground—a height to which few infective worm larvae will
venture.

So we reach the more surprising situation. The goat not only
consumes these 'damaging' worm larvae at a time when she is rising
to the peak of her condition, but also at times when they are
particularly hard to come by in the normal course of her grazing
habits. Finally the infestation occurs at times when the goat's diet is
changing from the hard foods of winter to the soft foods of summer,
and back again, from the soft foods of summer to the hard foods of
winter, and when a parallel change is taking place in the utilization
of the food nutrients, for the udder in summer, for body reserves in
winter. While it would be going too far to suggest that a heavy worm
infestation could have an effect that was actually beneficial, there
seems to be some indication that worms play some role in carrying
out these seasonal changes.

In any case, this is an academic point, for, in practice, total
wormlessness is a more or less unattainable ideal. The traditional use
of garlic (which is also a blood purifier) as a deterrent keeps infestation down to reasonable levels. It is particularly useful as a preventive measure, given to susceptible goats after three weeks on the
pasture; to all the grazing companions of a heavily infested goat;
and to all goats on close-cropped summer pastures after a fortnight
of warm, wet weather. Give a wild garlic plant or tablet on alternate
days for a week, immediately after the morning milking—this is
vital with plants, to avoid tainting the milk.

For routine spring and autumn worming, thibenzole is generally
used, and is considered safe and effective. Give this in powdered
form, made up into a drench. Nilverm is an alternative, given orally
or subcutaneously. New drugs appear, and will be prescribed by
vets, especially where intensive stocking calls for more drastic
measures.

The usual parasites, and details of prevention and treatment, are
listed under 'internal parasites' in the following section. From this
introduction, it will have become clear that the specific advice which
follows is based upon a compromise between 'nature cure' methods
and orthodox veterinary practices. For the undiluted 'nature cure'
doctrine, consult an appropriate herbal (and if you can find a copy

of the late F. Newman Turner's *Cure Your Own Cattle,* so much the better); for pure orthodoxy, ring up the vet.

Acetonaemia

Symptoms: goat goes off eating concentrates; may continue to eat roughage. Milk falls, irregular cudding. Droppings become dark, covered in sticky mucus. No fever.
Occurrence: usually within 2–3 weeks of kidding. Goat is usually a heavy milker.
Prevention: correct feeding. Ensure enough energy.
Cure: cobaltized salt for 10 days; increase roughage; injection of glucocorticosteroid will stimulate appetite, ½ lb (227 g) flaked maize night and morning will probably help.

Anaemia

Not a primary disease, but a symptom of cobalt deficiency (pine), or of worm and fluke infestation.

Blindness (contagious ophthalmia)

Symptoms: one or both eyes watery; cornea becomes cloudy; white spot develops at cornea. Spot enlarges, becomes reddened and may rupture, causing loss of the eyeball.
Occurrence: germ probably spread by flies, but also carried from infected eyes to healthy eyes by the tips of flapping ears while goats are feeding at trough. Usually more than one case in a herd.
Treatment: opthalmic ointment or antibiotic powder in eye are commonly used. Best treatment would seem to be subconjunctival injection of an antibiotic—long-acting penicillin possibly best.

Bloat (blown, hoven, tympany)

Symptoms: tightly inflated flanks; misery; collapse.
Occurrence: on lush clover and grass (especially wet grass) pastures in spring and autumn, gorging on anything unsuitable and after raiding the food bin.
Prevention: feed hay and allow an hour's cudding before turning out. Re-seed pasture with 'Herbal Lea' Mixture (see p. 303).

Cure: *Raw* Linseed oil 6–8 fl. oz. (150–200 ml) for an adult goat, 2 fl. oz. (50 ml) upwards for kids. Walk the goat about, massage flanks. After this treatment wind is usually expelled from either or both ends and goat rapidly returns to normal.

Coccidia

See 'Internal parasites', Table 8, p. 236.

Colic

Symptoms: spasmodic pain in the digestive tract; goat half rises, sighs and groans, goes down again, suddenly stops eating and looks anxiously about; may be tympany of the left flank; increasing distress; leads to death if unchecked.
Occurrence: commonest in young stock; conditions favouring worm infestation; excessive feeding of concentrates, access to cold water immediately after concentrate feed; in kids, after swallowing bottle teat; after poisoning.
Prevention: correct feeding and constant access to water; feed the flock roughages together; hold the teat when bottle feeding.
Cure: ½ pint (0·3 litres) of linseed oil (for adults, less for kids), followed by 1 glass spirits in 2 glasses of water, repeated hourly until pain subsides.

Cuts and Wounds

All cuts but the very smallest are best stitched, as this reduces the healing time. Most cuts are contaminated to some extent, and this both delays healing and provokes the goat to attack the wound, further delaying healing. Antibiotics reduce the infection which accompanies contamination, and speed healing. Tetanus antiserum is also indicated, unless the goat is vaccinated. A local or general anaesthetic may be needed for the cleansing and suturing of large wounds.

Cuts on the udder or teat exposing the teat must be stitched. The act of milking after suturing may re-open the wound, and the use of hollow teat cannulae with removable plugs to let the milk drain should be considered. Intramammary penicillin injection (both sides) is indicated to prevent the development of mastitis.

Dermatitis

See Eczema.

Diarrhoea

Not a disease, but the sign of disease caused by worms; coccidia; enterotoxaemia; salmonella and other bacteria; poisoning; indigestion; overfeeding.

In kids, it may be due to irregular feeding; or to the use of unsterilized feeding bottles.

Eczema

Several forms of skin disease occur in goats, two of which have the appearance of eczema.

Contagious Pustular Dermatitis (Orf)
Symptoms: pimples about the nose and mouth, less often about the eyes, anus and hoofs, turning to watery blisters; then to sticky and encrusted scabs. Also on udder of suckling goat.
Occurrence: a local infection of sheep country, commoner in eastern districts. Young animals most frequently affected; but all ages liable on first contact with infection.
Prevention: home-bred stock in an infected district have natural immunity. Vaccine will give protection to others.
Cure: difficult. Dress with gentian violet.

Goat Pox
Symptoms: pimples turning to watery blisters; then to sticky and encrusted scabs on the udder. If scabs removed before 'ripe', weeping sores remain.
Occurrence: common, with a wide variation of severity, depending on the virulence of the germ and the resistance of the stock.
Prevention: scrupulous dairy hygiene and isolation of affected milkers, which should be milked last and the hands disinfected thereafter, will help to control the spread of severe outbreaks.
Cure: time and gentle milking. A cream with antibiotic and cortisone may help.

Enterotoxaemia

This is the worst goat killer.

Symptoms: They vary somewhat according to the strain of bacteria concerned; but are usually consistent in any one flock. Staggering gait and loss of motor control is one of the less common symptoms; sometimes the goat is blown; there is always extreme misery; and, almost always, peculiarly evil-smelling diarrhoea. Coma and death within 24 hours are the normal sequel.

Occurrence: ineradicable bacteria, indigenous to goats, and to all other domesticated grazing stock and their pastures, produce the poisons responsible, when conditions in the digestive tract deprive them of oxygen. In goats, a big feed of lush, wet grass, or of concentrates and water, or a real belly-stretcher of milk, all produce the airless pudding in which the bacteria start poison production. Goats quickly build up resistance to the poisons produced in small regular quantities; it is the sudden change of weather and diet in spring and autumn, and the accidental gorges of goats in mischief, that cause most trouble.

Prevention: correct feeding on a bulky, fibrous diet. Biennial treatment with vaccine gives protection.

Cure: on first acquaintance, enterotoxaemia is usually fatal before diagnosed. Once the characteristic symptoms have been seen, they are more easily recognized when they recur.

Immediate resource to the hypodermic syringe will usually save the goat. Two injections, each of 10,000 units of penicillin, four hours apart, or a single injection of 300,000 units of slow-release veterinary penicillin, is likely to be successful; sulphamezezine (dose according to weight) gives good results. Whatever treatment is to be used, it must be kept in permanent readiness for immediate use. Delay proves fatal.

Ergot

See Poisoning.

Fluke

See Internal parasites, Table 8, p. 236.

Foot-and-mouth Disease

Symptoms: sudden lameness; small, dribbling blisters on the tongue, inside of lips and on the palate; also where hair joins hoof and between the claws of the hoof. Can be confused with contagious pustular dermatitis.
Occurrence: rare. Inform the police. Don't seek advice.

Foot rot

Symptoms: slight and increasing lameness in one or more feet; hoof uneven, with stinking matter oozing between the outer horn and inner soft structures.
Occurrence: among goats on sodden pastures or floors; but the disease is either a regular visitor, or an infected sheep or goat introduces it. The bacteria concerned can remain latent for long periods in the hoof but cannot survive a week on the ground.
Prevention: dry floors and the avoidance of sodden pastures; regular exercise on hard ground. Regular trimming of feet, vaccination with foot-rot vaccine.
Cure: pare the foot level; remove loose horn and, very gently, as much dead matter as possible. Soak all four feet for 20 seconds in a 'Dettol' solution (4 tablespoons to 1 pint [0·6 litres] water). Repeat daily till sound. Then, once every 10 days for a month, dip the feet briefly in a copper-sulphate solution or proprietary foot-rot wash. Keep the infected feet, if not the goat, out of contact with the communal pasture, using doll's Wellington boots, perhaps. A 10 per cent solution of formaldehyde is quite effective as a foot bath. Even a small amount in a tin can, in which the feet can be immersed above the hair line one at a time, is sufficient. Goat should then be stood on concrete or wooden floor until solution is dry. Repeat weekly for 3–4 weeks.

Fractures

A fracture (if skin not broken) below elbow or fore leg below hock or hind leg can usually be treated by P.O.P., depending on value of goat. Surgery may be successful in other cases. Consult vet.

Table 8: Internal parasites

	Round worms of digestive tract	Lung thread worms	Lung hair worms	Tape worms	Liver flukes	Coccidia
Symptoms:*	Scour, soft lump under jaw. Alternatively, anaemia, wasting, no scour.	Harsh cough and sticky discharge from nose. Wasting.	Dry cough. No discharge. Poor condition.	Usually mild. Wasting, rickets, tetany, occasionally.	*Acute form:* Dullness, distended abdomen painful on pressure at junction of ribs and breastbone. Death. *Chronic:* Dullness, anaemia, wasting, pale gums, soft swellings under jaw and abdomen Only slight scour.	Continuous or recurrent scouring, often profuse and blood-stained. Abject misery, no appetite. Death may occur after 5 days' illness. Relapse 2 weeks after apparent recovery.
Occurrence:	All year round. All ages. Severe only when grazing short grass or arable forage crops.	Severe in kids, mild in adults. Wet pastures and weather. Autumn.	All year. Adults. Wet grass.	March to October Mainly in kids.	August to January. All ages. Presence on pastures of mud snail. Wet summers.	May to October, All ages, worst in kids.

Other hosts:		Sheep and deer.		Sheep	Sheep, cattle, deer and rabbits.	Sheep and wild birds.
Prevention:†	Rotational grazing—1 week on, 3 to 4 weeks off. Alternate stocking. Routine garlic.	Prevent access to pasture in dull, wet weather. Alternate stocking	Dress land in summer with 20 lb (9 kg) ground copper sulphate mixed with 80 lb (36·2 kg) sand per acre. Alternate stocking.	Bottle feed kids. Alternate stocking. Routine garlic.	Drain and dress wet patches as for lung hair worm. Dose all liable stock with hexachlorethane in October. Run ducks on wet pastures to eat out the fluke.	Generally good hygiene.
Cure:	Thiabenzole; Nilverm; Panacur			Copper sulphate and nicotine	Hexachlorethane	Sulphamezathine; Panacur
			or, in all cases, drugs prescribed by veterinary surgeon			
Diet Supplement:‡	Seaweed meal and B vitamins	Garlic and seaweed meal			Seaweed meal, lime-rich foods and Vitamin D	Vitamin A

Notes to table 8.

* It happens more often than not that a goat heavily infested with one type of parasite is infected with several types, so symptoms may be mixed. Identification of the dominant parasite concerned is seldom possible unless symptoms are considered in conjunction with the conditions of occurrence. The examination of dung in a veterinary laboratory may help, on occasion, to diagnose the parasite; but in well-fed goats, serious infestation with few worm eggs in the dung is a common concurrence.

† Goats, like other ruminants, are in constant contact with the infective larvae of internal parasites which exist on the herbage, and habitually carry a small worm burden, which is harmless and part of the goat's way of life. Serious infestations arise from bad management—either in confining the goats to overstocked and unsuitable pastures; or in breaking down their natural resistance by bad feeding, bad housing and over-exploitation. Preventive measures suggested here are designed to minimize the consequences of bad management. 'Routine garlic' means giving each goat a garlic tablet or a wild garlic plant on alternate days, or every day, immediately after milking in the case of milkers. This measure enables goats to tolerate exceedingly bad pasture management with impunity.

‡ This column reflects current knowledge or knowledgeable guesses concerning the substances of which the various parasites rob their hosts. These are chiefly minerals and vitamins. Seaweed meal is advocated as a mineral supplement because the most innocuous and palatable available. For vitamin B, wheatgerm meal, Bemax, Phillip's poultry yeast, and Marmite, are good sources. For vitamins A and D, either the dry vitamin supplement sold by leading agricultural chemists or the concentrated synthetic preparations, designed for infant feeding, are recommended—e.g. Adexolin. Neither cod-liver oil nor natural sources of vitamin A are suitable, as the former is indigestible and the latter require the efficient functioning of a damaged liver.

Gas gangrene

Occasionally infects a deep penetrating wound, producing a foul-smelling bubbly discharge. Death may follow in a few days. At an early stage, antibiotics and serum may control the infection. Vaccination against clostridia generally may prevent this disease.

Goat pox

See Eczema.

Hoven

See Bloat.

Johne's Disease

Symptoms: loss of condition; occasional scouring, becoming more frequent with bubbles of gas in the droppings; appetite usually good; emaciation; weakness; death; may take several months to run its course.
Occurrence: infection may be picked up from cattle or sheep.
Prevention: management to prevent spread of infection.
Cure: none.

Lactation Tetany

Symptoms: anxiety; uncontrolled movement; staggering; collapse; convulsions—all in rapid succession—death if not immediately checked.
Occurrence: commonly when put out on to pastures in the flush of spring; also in autumn towards height of breeding season.
Prevention: correct feeding and management: especially, provision of magnesium at times of likely deficiency.
Cure: immediately inject about 100 ml of 25 per cent magnesium sulphate solution subcutaneously in loose skin behind elbow. A veterinary surgeon *might* inject a weaker magnesium solution intravenously.

Lice

Symptoms: skin irritation; rubbing, scratching etc.; bald patches; lice seen on skin.
Occurrence: very contagious; common in goats in poor condition, especially those with little access to open air, and to rainfall.
Prevention: regular treatment.
Cure: B.H.C. dusting powder; or wash weekly for 3–4 weeks with Derris or B.H.C.; or single bath with organophosphorous insecticide. This will control ticks and mange mites as well.

Louping Ill

Symptoms: dullness and fever; followed by tremor, muscular spasm and bewildered gait; often by collapse, paralysis and death.
Occurrence: on tick-infested pastures in Scotland, North of England and North Wales during the period of tick activity—April and August–September. Home-bred young stock are only slightly affected, seldom developing the nervous symptoms; home-bred adults are very seldom affected at all; newcomers of all ages are liable to serious infection during the first tick season.
Prevention: home-bred stock develop natural immunity. In-bought stock from a tick-free area should be inoculated with 5 ml louping-ill vaccine in July and March of their first year.
Cure: shelter, and quiet and careful nursing, will often save the patient. The presence of a congenial companion in the sick bay is a help.

Mange Mites

See Lice.

Mastitis

The goat is liable to all forms of mastitis that affect the cow; and also to black garget of sheep. Several different diseases with different causes are grouped under this head. Only in a minority of cases is mastitis in goats associated with specific bacteria; many cases are not associated with bacteria at all; some forms are highly contagious to all goats; others are contagious only to goats in an unhealthy condition; others again are not infectious at all.

Acute Mastitis
Symptoms: misery; udder hot, hard and very tender; milk clotty and often blood-streaked; appetite lost; pupils of eyes narrowed to slits. In the worst cases, the udder may putrefy and slough away, and the goat die in high fever.
Occurrence: in goats fed a concentration diet, especially during the first fortnight after kidding; and following injuries to the udder. Highly contagious to goats similarly fed.
Prevention: correct feeding and management; let kids suck for first four days of lactation.
Cure: penicillin.

Sub-acute Mastitis
Symptoms: as for acute mastitis; but appetite near normal, and less distress.
Occurrence: as for acute mastitis; but at any time during lactation.
Prevention: as for acute mastitis.
Cure: as for acute mastitis.

Chronic Mastitis
Symptoms: lumpy udder; occasional clots in the milk; occasional off-flavours; no notable discomfort or distress.
Occurrence: as for acute mastitis.
Prevention: as for acute mastitis.
Cure: as for acute mastitis.

Summer Mastitis
Symptoms: dejected or anxious expression; slitty eyes; lumpy udder, which may develop into acute inflamed udder, which putrefies or shrinks and hardens into a dead lump. Often fatal.
Occurrence: mainly in goatlings or older kids during summer; occasionally in dry goats; only affecting flocks in which large quantities of concentrates or concentrated succulents are fed to young stock.
Prevention: correct feeding and management.
Cure: as for acute mastitis.

Black Garget
Symptoms: as for acute mastitis; but udder rapidly becomes black and gangrenous, either sloughing away or hardening into a dead lump. Often fatal.
Occurrence: occasional in all goats after a wound on the udder, particularly a penetrating wound; in goats suckling a lamb.
Prevention: remove barbed wire from fences; control dogs; bottle-feed orphan lambs.
Cure: a job for the vet, who may be able to get the antiserum in time; and otherwise may succeed with antibiotics, or iodized oil suffusion. While awaiting the vet's arrival, carry out hot and cold fomentation, massage and frequent stripping.

Metritis
Inflammation of the womb.
Symptoms: slowly developing feverishness to point of acute misery and collapse, usually with increasingly foul-smelling discharge from the vulva.
Occurrence: not before kidding, unless the kids are dead. Within a few days of kidding, especially if kidding is assisted, or cleansing retained, or incomplete kidding (retained kid).
Treatment: antibiotics.

Milk Fever
Symptoms: incoordination; staggering; collapse and inability to rise.
Occurrence: maybe a month before kidding to a week after kidding; but usually within 2 days of kidding.
Prevention: correct feeding will not prevent this disease.
Treatment: 200 ml of 20 per cent calcium borogluconate at blood heat subcutaneously. Inflation of the udder is unnecessary.

Ophthalmia
See Blindness.

Orf
See Eczema.

Pink Milk

A trace of blood in the milk is common shortly after kidding, especially in first kidders, and is due to the bursting of small blood vessels in the udder. If it is not accompanied by other symptoms, this is innocent. If the condition persists it indicates lack of calcium (or an excess of phosphates, i.e. concentrates) in the diet.

Pneumonia

Symptoms: signs of fever, i.e. depression; off food; standing with back arched; coat on end; breathing faster than normal, may be laboured and with grunt.
Occurrence: after stress.
Cure: antibiotics—tetracycline, etc.

Poisoning

Generally speaking, goats are fairly immune to semi-poisonous plants if on free range with plenty of variety; but there comes a time when forage is scarce, or a gate left open, so they gorge on something unsuitable with dire results. Prompt action is absolutely essential: decide what the trouble is and tackle it immediately. *Poisonous Plants*, published by H.M.S.O., is useful for identification.

The principle is the same in all cases. First remove the goat from the poison, then the poison from the goat. Having removed the goat from the poisonous shrub, food bin or whatever; house her in a warm place covered with a rug; and, if extremities are cold, massage and surround with covered hot-water bottles. Give a purge or emetic, generally Epsom salts. The next stage is the antidote to neutralize the poison, followed by stimulants, or whatever is necessary to heal the damaged tissues. Chlorodyne should be at hand, and given if pain is acute—$\frac{1}{2}$ teaspoon for an adult goat. Treatment of shock is vitally important; extremities tend to get cold while the system works overtime ridding itself of the poison. Strong, sweet, black tea or coffee stimulates and counteracts shock—but it must be strong.

A linseed-oil drench is frequently advisable as a follow-up measure to offset the caustic effects of Epsom salts. Dosage varies from 6 fl oz to $\frac{1}{2}$ pint (170 to 284 ml) or more, depending on severity of poisoning and size of goat. Small soft-drink bottles are convenient for administering this.

After-treatment consists of bran and molasses mashes; no concentrates at all; hay; fresh water offered or changed frequently.

Special indications for particular vegetable and mineral poisons, and for overeating of concentrates in a raid on the food store, are as follows.

Acorns (unripe)
Treatment: ½-pint (0·3 litres) linseed-oil drench; laxative food.

Arsenic
Symptoms: slobbering; scouring; thirst; attempts to vomit; onset rapid.
Occurrence: found in weedkillers, sheep dips, etc.
Treatment: 6 oz (170 g) Epsom salts in ½ pint (0·3 litres) water as purge; follow with linseed-oil drench and white of egg in milk.

Azalea
Treatment: as Rhododendron.

Beet Leaves
Treatment: suspension of ground chalk in milk as antidote.

Concentrates
Prevention: it is a golden rule that bins or bags containing concentrates should never be accessible to stock. A goat is exceedingly artful, and will even manage to undo many types of catch before pushing up the lid of a bin. There should never be the least possibility of this happening.
Treatment: speed is vital. Purge with 6 oz (170 g) Epsom salts in ½ pint (0·3 litres) water; get this down somehow. Withhold water, which swells grain and makes things considerably worse. If the poison is grain, the vet will give an intravenous injection of Parentrovite (vitamin-B complex). Treat for shock (rugging, warm stabling); follow later with smallish linseed-oil drench.

Conifers (except Yew, q.v.)
Treatment: Epsom-salts purge, size of dose according to severity of poisoning.

Ergot
Symptoms: lowered condition; cold and loss of sensation in the extremities.
Occurrence: among goats confined to pastures on which ergot, a fungus growth on flowering grasses and cereals, is prevalent.
Prevention: keep goats away from such pastures.
Treatment: none is effective.

Hemlock (roots in winter)
Treatment: strong, sweet, black coffee or tea with 1 fl oz (28 ml) spirits as stimulant; keep goat moving.

Herbicides and Pesticides
Symptoms: typically, sprays cause violent black scour and attempts to vomit.
Occurrence: many goats have been killed when browsing near the hedge in a field next to one being sprayed.
Treatment: 6 oz (170 g) Epsom salts in ½ pint (0·3 litres) water as purge; later, smallish linseed-oil drench. After such poisoning, it is vital to keep the goat off any form of greenstuff for several weeks, or even months, until all after-effects have disappeared. (See also Arsenic.)

Laburnum
Treatment: as Hemlock.

Laurel
Treatment: as Rhododendron.

Lead
Symptoms: colic; intense pain; goat grinds teeth and slobbers; often followed by delirium and blindness.
Occurrence: found in paint, linoleum, roofing felt, etc.
Treatment: there is one specific and highly efficacious antidote, which should be given in cases both where symptoms are apparent, and where poisoning is merely suspected. This is calcium disodium versinate solution, made by Riker; it is given intravenously.

Mangold Leaves
Treatment: as Beet leaves.

Pesticides
See Herbicides and pesticides; also Arsenic.

Rhododendron (and similar shrubs)
Symptoms: vomiting; spitting out greenstuff; pain.
Treatment: there are two alternatives. Vets recommend 6 oz (170 g) Epsom salts in ½ pint (0·3 litres) water; 1 pint (0·6 litres) liquid paraffin (for demulcent effect); very strong, sweet black coffee.

The usual goatkeeper's treatment is: first induce further vomiting with 2 tablespoons melted lard and 1 tablespoon bicarbonate of soda, shaken together in a warm bottle and carried to the goat house in a jug of warm water to stop it solidifying. Follow 1 hour later with 1 cup molasses and 1 tablespoon bicarbonate; 2 hours after this, strong tea or coffee as above.

Sheep Dip
Often contains arsenic, *q.v.*

Snake Bite (Adder)
Symptoms: shock; if snake bite suspected, look for swelling and two small punctures. Adder bite is seldom fatal, except on udder.
Occurrence: the adder or viper is the only poisonous snake in Britain; from springtime onwards it likes to lie sunning itself, usually on piles of stones, flat rocks or dry banks. Black-and-silver colouring (brown just before casting skin) with V markings all the way down back; quite unlike beige, unmarked, non-poisonous slow worm. Goats less likely to be bitten than dogs since, on finding a snake, they stand still, and the shy snake slinks away.
Treatment: rub permanganate-of-potash crystals into wound; treat for shock. (This may also save a dog, if done within 20 minutes or so.)

Weedkillers
See Herbicides and pesticides; also Arsenic.

Yew
This dangerous plant can kill within minutes.
Treatment: unlikely to be effective; but try usual poison routine, with extra-large linseed-oil drench (up to 1 pint [0·6 litres] for a large goat) and 1 fl. oz. (28 ml) spirits. Keep goat moving; follow up

with molasses and strong tea or coffe as for Rhododendron, but omitting the bicarbonate of soda.

Pregnancy Toxaemia

Symptoms: dullness, stupidity; loss of appetite; collapse; death—all within 12–48 hours.
Occurrence: occasionally in goats on all standards of nutrition whose diet fails to improve in quality during the last six weeks of pregnancy; lack of exercise increases liability.
Prevention: correct feeding and management, with special reference to steaming-up and regular exercise.
Cure: inducing abortion with cortisone may save goat. Nothing guaranteed, but 2 oz (56 ml) glycerine mixed with 2 oz *boiling* water, given as drench twice a day, works well with ewes.

Ring Worm

Symptoms: round, dry, scabby patches appear on the skin. Contagious to humans; infections can remain in buildings and re-infect.
Prevention and cure: vet will supply the feed additive Fulcin, which should be fed to all goats.
   Scrub out sheds and boxes with washing soda in strong solution; treat wood with creosote.

Tetanus

Symptoms: tense, anxious posture, head held up, neck outstretched, ears pricked, top lip retracted in snarl, tail elevated; slight tympany on left; legs rigid, like rocking horse.
Occurrence: following any wound, whether deep or not, if contaminated by the tetanus germ—even following disbudding a kid. Occasionally spontaneous, with no known injury.
Prevention: either tetanus antiserum at time of injury, or incorporate tetanus in the vaccination programme.
Cure: massive doses of tetanus antiserum to counteract the toxin, and penicillin to kill the germ—recovery doubtful.

Tetany
See Lactation tetany.

Ticks
See Lice.

Toxaemia
See Pregnancy toxaemia.

Tympany
See Bloat.

Worms
See Internal parasites, Table 8, p. 236.

Foot Trimming to Avoid Infection Malformation

The dairy goat lacks the exercise on rock which keeps the wild goat's feet in shape, and it becomes essential to cut away the outer horn and very carefully trim the hoof so that it stands squarely on a flat surface. Feet should be attended to every four weeks, starting when a kid is about six weeks but omitted in the last two months of pregnancy. All too often feet are neglected, the outer rim turns over, grit and mud accumulate and lead to infection, which can even be transmitted to the rest of the herd. Misshapen feet are sometimes impossible to reclaim after a long period of neglect, so this regular trimming is a necessary routine exercise. If started early in life, with a sharp knife or foot-rot shears, and provided that care is taken over the job, which should be done when the hoofs are soft after a period on wet pasture, the goat accepts the operation, instead of fighting against it. There is a right and a wrong way to tackle it, and a demonstration by an experienced goatkeeper is the sensible approach.

ALTERNATIVE MEDICAL TREATMENT

A number of the most important goat diseases—e.g. mastitis, enterotoxaemia and some forms of pneumonia—may also be treated on the basis that these diseases are primarily due to the accumulation of toxins in the organs of the goat, the presence of which provide the conditions necessary for the action of specific disease bacteria. Garlic treatment has been recommended to aid the removal of such toxins, as this method has been most widely used for the purpose.

Plate 15. This pleasant British Alpine milker would stand better for her photograph if her hooves were trimmed by regular exercise on hard ground. 'Dutch clogs' are apparent on her near hind hoof, but the 'back on her heels' stance is typical of the goat whose feet are maintained by the paring knife instead of exercise. She was bred by Mr Egerton and gave 4,200 lb in 365 days.

Now the goat is capable of a daily throughput of digestible nutrients, per 100 lb (45·4 kg) of bodyweight, twice as great as that of cow or sheep. She is therefore liable to auto-intoxication at a rate twice as high as that of the cow or sheep. She must either prove an exceptionally unhealthy animal, or be provided with a compensatory detoxicating mechanism. It is significant, therefore, that the goat suffering from pernicious anaemia resulting from cobalt deficiency requires four times as much cobalt to cure her as the amount required by an anaemic sheep of the same size. Pernicious anaemia is not directly due to lack of cobalt in the blood of the goat, but lack of vitamin B$_{12}$, which is synthesized by the rumen bacteria from cobalt and other raw materials. Vitamin B$_{12}$, in alliance with vitamin B$_1$, and other substances of a water-soluble nature, are known to act in a way which suggests that they are detoxicants; human need of them, for instance, is in proportion to carbohydrate intake in many cases.

It appears reasonable to suggest that the high rate of auto-intoxication of the goat, when fed to the limits of her appetite, is naturally balanced by a high capacity for vitamin synthesis; that the main natural detoxicant of the goat is this combination of B vitamins and allied substances with B$_{12}$ and B$_1$ playing a major role.

On a fibrous diet, B$_1$ requirements will always be met unless there is a heavy worm infestation which grabs the vitamin before the goat gets it—the substance of parasitic worms is one of the most vitamin-rich natural foods available. But the supply of cobalt for the manufacture of vitamin B$_{12}$ is by no means assured. Areas where the cobalt content of herbage is inadequte for sheep are already known to be extensive, and are still being mapped; heavy cropping and modern methods of husbandry tend to lower the natural cobalt content of fodder.

It may well be that the use of doses of these vitamins and of B$_{12}$ in particular will be found to provide a more natural rapid and effective detoxicant than garlic—which is not without its disadvantages. The vitamins cannot however be provided through the mouth, as the greater part of them is then decomposed before being absorbed; they must be injected.

There are no adequate grounds for recommending this method of treatment in a work of this kind; but there seem to be sufficient grounds for an interested veterinary surgeon or goatkeeper to try these methods in cases where other methods fail to give satisfaction.

There is no doubt that a simple herbal remedy may succeed when orthodox treatment fails. The dock-and-elder brew for mastitis has achieved considerable success. Take a handful of each, shred finely, put in an enamel (not aluminium) pan and add boiling water. Cover with a tightly fitting lid and leave overnight. Next day, strain; at milking, with a wrung-out cloth, slop the liquid all over the udder, and then massage. This should be repeated every milking till the mastitis clears up; renew the brew daily. When leaves are not in season, roots can be used; but cover these with cold water and boil two or three minutes before leaving to infuse. Wood sage is an alternative treatment. Bramble leaves are much liked by goats, and a useful antidote to fresh spring grass. Ivy—without berries—will tempt a goat with poor appetite after a digestive upset.

There are numerous other herbal remedies, of varying efficacy; but this book is no place for them. In her *Herbal Handbook for Farm and Stable* (Faber), Juliette de Baïracli Levy enlarges on the effectiveness of herbs. Both she and the late Newman Turner (in *Cure your own Cattle*) place great emphasis on fasting. With the digestion at rest, all energy is concentrated on fighting the disease. Put a goat on a forty-eight hour fast, giving only water and garlic. She emerges fit and hungry. Never mind the drop in milk, which is likely not only to rise quickly, but even to increase. Orthodox treatment invariably urges encouraging a sick animal to eat, when its own instinct tells it to lie up quietly without food. A prolonged fast calls for the addition of garlic enemas.

Table 9: First-aid kit for the goat house

A sharp knife	Linseed oil
2 in and 1 in (5 and 2·5 cm) Prestoband bandages	Drenching bottle
	Hypodermic syringe
Adhesive plaster	Enamel bowl
Surgical gauze	Garlic tablets
Cotton wool	Tree-bark food
Surgical thread and needle	Bicarbonate of soda
Boiled brine (or Dettol)	Veterinary penicillin
Tincture of iodine	Sulphamethazine
Acriflavine emulsion	Chloromycetin (aerosol spray)
Hydrogen peroxide	A goat blanket

The normal temperature of the goat is 102·5 to 103°F (39·1 to 39·4°C).
The normal pulse rate of the goat is 70 to 95.
The normal respiration rate is 20 to 24 per minute.

Chapter Eleven

Milking Practice and Dairy Produce

Whether the end product of our goat dairy be fresh milk, cream, butter or cheese, the prime requirement is a supply of clean milk of regular composition.

The cleanliness of goats' milk must inevitably be judged by the standards obligatory on cows' milk. There are, in effect, two standards: 'certified' and 'pasteurized'. If goats' milk is to serve the quality market with particular reference to invalids and young children, there is only one standard to be considered—the best.

In seeking to attain the standards laid down for 'certified' cows' milk, the goatkeeper has two significant advantages and one serious disadvantage.

The advantages are, first, that the milk of the goat naturally has a lower bacterial content, as it comes out of the teat, than the milk of the cow. In the second place, the conditions under which goats' milk is produced are not subject to the same legal controls as are imposed on cows' milk production; it is therefore permissible to use chemical methods of sterilizing goat dairy equipment instead of the very much more laborious and expensive steam-sterilizing equipment which is obligatory for cow dairymen.

The disadvantage the goatkeeper has to face lies in the character of goat dung. Certified milk standards demand that there should be no coliform bacteria present in a sample one-tenth of a cubic centimetre of milk. Coliform bacteria of the type most likely to find their way into goats' milk are not particularly dangerous to humanity; but they are capable of multiplying in the milk at a higher rate than most other forms of bacteria, thereby souring the milk and producing off-flavours very rapidly. Coliform bacteria find their way into milk mainly on the vehicles provided by minute particles of dung dust. Cows' dung is far from dusty, and if the cow byre is richly charged with dung dust at milking time, it is a very dirty and sloppily managed byre. So the coliform bacteria test is a sound criterion of clean cows'-milk production.

But goat dung is exceedingly dry and dusty; moreover the goat's need for a warm bed and some freedom of movement makes it impossible to achieve with her the clinical cleanliness of the concrete cow byre and the same scrupulous isolation for her dung. Even if we put the goat on some form of insulated and impervious floor, her dung pellets hop about like a packet of spilled peas, here, there, and everywhere—under her feet to be trampled to dust, and on to her bed to contaminate her coat.

The cow, who stands belly-deep in the river for half of the summer day, for preference, has no objection to a bucket of water splashing round her legs when a cow pat goes astray in the byre. But such hygienic procedure in the goat shed would lower yields all round.

We can therefore assume that the goat house and goat, at the best of times, will have a far higher content of dung dust and coliform bacteria than any decent cow or cow byre would own.

Some diligent goat dairymaids counter this difficulty by milking into a milk strainer set over the pail. Unfortunately, this precaution seldom has the desired effect. The dust and hair, to which the coliform bacteria are attached, accumulate in the strainer, while the coliform bacteria themselves are power-sprayed off their vehicles by the jet of milk, through the filter and into the milk. So long as the bacteria remain attached to hair and dust there is a reasonable hope of extracting the greater part of them in the dairy filter; once they are separated from their vehicles, they are in the milk for good or ill. The use of hooded milking pails is a help in reducing the amount of hair and dust which comes into contact with the milk; but they are more difficult to sterilize and, at best, they do not bridge the obstacle to hygiene constituted by the goats' dusty dung.

There can be no half measures. No goat house is suitable for producing goats' milk to 'certified' standards. A milking parlour, which is a useful aid to the production of clean cows' milk, is a prime necessity for the production of clean goats' milk. The milking parlour need not be expensive, large, or elaborate; but it must be separate from the goat house, and it must be washable.

254

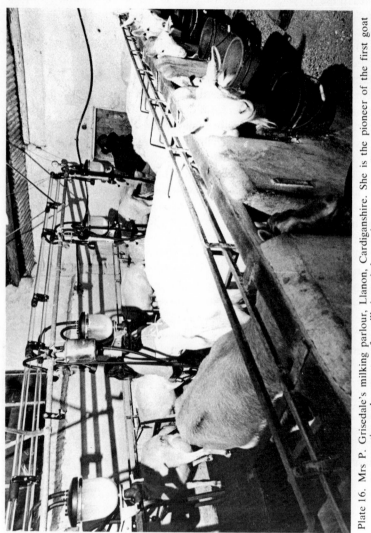

Plate 16. Mrs P. Grisedale's milking parlour, Llanon, Cardiganshire. She is the pioneer of the first goat co-operative and promoter of a milk (spray) dryer scheme.   (Photograph: Michael Murray)

Her Majesty's Stationery Office publish a leaflet on cow milk parlours. Most of the material therein is directly applicable or readily adaptable to goat milk parlours. The main points are: an impervious, self-draining floor, coved at its junction with the walls; inpervious, washable walls to at least 4 ft (1·25 m) height; general construction to avoid horizontal shelves and ridges which will trap dust; facilities for holding, and perhaps, feeding, the goat while being milked. Corrugated iron is too noisy, heat-conducting, and perishable to be recommended, but almost any other material is suitable for the main structure. Fig. 20 shows a sample goat milking parlour, designed to provide facilities for machine milking. Most types of existing cow milk parlours are quite suitable for use by goats; only the tandem type presents some difficulties.

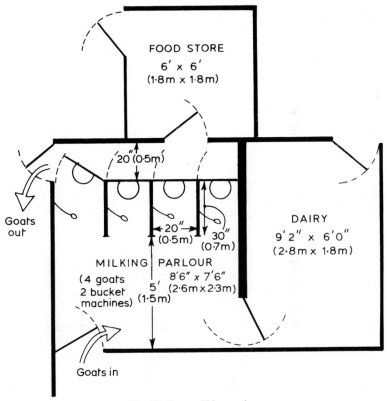

Fig. 20. Goat milking parlour

The goats will, of course, be brushed down before entering the milk parlour, but their coats will still be rich in dung dust and coliform bacteria. As the goat offers less room for the milker to manoeuvre than the cow, her coat suffers more disturbance in the process of milking, so that a relatively clean goat would release more dung dust at milking time than a relatively dirty cow.

There are two ways of dealing with this problem: one or other or both of them must be adopted unless we are going to set ourselves a serious handicap to clean-milk production. Firstly, there is no reason why the dairy goat should not be clipped, at least over the area of its coat which is disturbed during milking. Anglo-Nubians of many strains are content with a reasonable modicum of hair; and there is no excuse for other breeds to be flapping their dung-dusty locks around in the milking shed. The whims of the 'fancy' and the show ring must never be allowed to frustrate clean milk production: if the goat of the show ring is to participate in commercial milk production and retain her natural coat, a Hoover 'Dustette' will have to become permanent goat-house equipment! And why not?

Secondly, goats can be milked by machine, by which means very little dust reaches the milking pail. Goats respond exceedingly well to machine milking, but there are some snags. The weight of the standard teat cluster—even when reduced to two teats—is excessive for the goat's udder. The goat milking machine is accordingly designed with a claw piece that rests on the floor. With no weight at all to hold it down, the teat cluster has a tendency to climb up the teat in the later stages of milking and, in the case of a badly formed udder, to swallow a proportion of the udder. However, given a goat with a clear line of demarcation between teat and udder and proper supervision, machine milking is perfectly satisfactory. The sphincters on the goat's teat are less powerful than those on the cow's, and a lower vacuum setting (as advised by the makers of the machine) is required. For an average goat, giving 4 pints (2·3 litres) at each milking, the total milking time is four minutes per goat, including tying and udder washing, provided a machine is used. In this period a goat can also consume $\frac{3}{4}$ to 1 lb (340 to 454 g) of concentrates. One man, with the one machine which he can effectively control while feeding and udder washing, can put through the milking parlour the twenty or thirty goats in about two hours. With two machines and an assistant in the milking parlour, the job can be done in an hour or slightly less. Six to seven minutes per goat is needed by the hand milker.

The milking machines adapted for goats are of the bucket type. There would appear to be no technical difficulty of much importance in the way of adapting recorder units for goats, if the demand arose. But where a flock kept under non-intensive conditions is producing less than 6 pints (3·4 litres) average peak yield, and for flocks of less than ten milkers, a hand milker can get through the job more quickly than he can set up the milking machine, do the milking and clean the machine. Specialist goatkeepers do well to consider that many milking machines on small dairy herd farms absorb as much labour in being cleaned as they save at milking; their existence is often due solely to the fact that farmers' wives do not draw the statutory minimum wage.

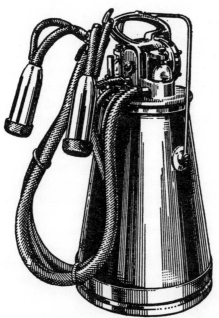

Fig. 21. Gascoigne goat unit

But the dairy farmer with a bucket plant has only to exchange the standard cow teat cluster for the special goat cluster, and reset the vacuum control, to reap the advantages of machine milking of goats with no great alteration in wash-room labour. In such case, irrespective of time saving, the goat machine may be valuable simply to eliminate dung dust.

In organizing the milking, routine labour economy is most important. When all the labour resources of the farm are mustered at the milking, it is most expensive to have them and the machines standing idle while Matilda finishes up her corn ration. Cattle standards can be misleading:

A Friesian cow takes 1·91 minutes to eat 1 lb (454 g) of grain.

A Jersey cow takes 2·39 minutes to eat 1 lb of grain.

A 100-lb (45·4 kg) goat takes 5·04 minutes to eat 1 lb of grain.

A 150-lb (68 kg) goat takes 4·23 minutes to eat 1 lb of grain.

Small goats and goats fed large concentrate rations should be fed before or after milking if they are to be milked by a machine, which seldom requires more than four minutes per goat.

In only one other respect does the production of clean goats' milk present special difficulties: the possible presence in the near vicinity of a male goat. It is absolutely essential that no taint of 'billy' come near the milking shed. Billy owners' sense of smell becomes mercifully blunted in course of time. Unless there is a rigid routine as to the times at which the billy's needs are attended to, and as to what happens to the hands and garments that approach him, his taint will find its way through the best of intentions.

In other respects the factors governing the production of clean goats' milk are fully covered by the massive literature on clean-milk production offered to the dairy cattlemen. It would be a waste of space to reprint them here. No one attempting to sell milk should fail to study one of the many cheap and excellent leaflets on dairy technique.

Those who produce milk for sale and follow the advice given above are earnestly recommended to submit regular samples for analysis. If the efforts to eliminate dung dust are thoroughly carried out and successful, there is every reason to suppose that local 'clean milk' competitions should never be won by a cattleman if there is a goatkeeper among the competitors.

For domestic use, milk produced by one's own goats is none the worse for the presence of a few coliform bacteria per cubic centimetre. We swallow them by the million whenever we clean the goat house. But since those who drink the milk of their own goats will seldom study dairy textbooks, the very elements of hygiene may—and often do—escape them.

Never use for human consumption any of the milk from any goat with any disease. Strain and cool milk as soon as possible. Wash all milk pails, jugs, dishes, etc., as follows.

(1) As soon as possible, fill, cover and rinse them in cold water.

(2) Fill, cover and wash them with warm water in which dairy detergent has been dissolved (but NOT IN HOT WATER—the temperature of the water should be no higher than that used for the baby's woollen vest or for best nylons).

(3) Rinse in plenty of warm or cold water.

(4) Scald with boiling water, and DO NOT DRY WITH A CLOTH, but stand upside down to dry on a RACK, not on a shelf, so that the air can get in and the moisture can get out. Unless the milk pails, etc., are kept in a sterile dairy, if you want to be particular, or there is a disease in your house, scald the pails again immediately before use; alternatively, put them for a quarter of an hour in an electric oven set at 250°F (121°C).

It is worth while sterilizing the cans in the oven periodically, for the smell of the can on removal from the oven is a good working test of cleanliness. Should the result of this test prove shocking, or should the keeping qualities of the milk deteriorate, or your regular visitors stop taking milk in their tea, do not waste your time and energies but buy a 'milk-stone' solvent from a dairy supply firm. Whenever a trace of milk remains in a can which comes in contact with hot or scalding water, an invisible film of albumen is formed on the surface of the can. This film feeds bacteria, and taints the milk; but so gradually that the regular consumer is liable to ignore the change. This film can be removed effectively only with the 'milk-stone' solvent.

For cheese making on a considerable scale, it is essential—and for fresh milk retailing, it is highly desirable—that the milk produced should be reasonably constant in composition. Some seasonal fluctuations in composition are inevitable; but day-to-day fluctuations can be controlled by regular milking routine and correct milking practice. In addition, both these factors have an important influence on yield.

Milking should be at regular intervals—as near twelve hours as may be—otherwise, the milk drawn after the longer interval will be relatively low in butter fat.

The actual routine of milking will depend upon the standards of hygiene required and individual circumstances. But it must be a rigid, orderly routine, involving the minimum of fuss and bother.

The goat requires at least five minutes' warning that she is about to be milked, in order that the operative sphincters and valves in the udder may release all the milk available when required. During the ensuing five minutes, there must be no distracting novelty in procedure, or the 'let-down' will be held up. There must be no undue delay once the process is under way, for if the goat is disappointed in the expected moment of her milking, the operative sphincters will impatiently close again.

This mechanism must be remembered when arranging the routine for milking the first goat: the fact of the first goat being milked is adequate warning to the next on the list; but the first goat is entitled to some warning herself. Where udder washing and drying is practised, this procedure will constitute fair warning. Where it is not practised, some routine which will command the goat's attention must be incorporated in the programme.

The goats must be milked in a set order. In a flock which enjoys some measure of social organization, the order should be that of the social hierarchy—an order which a little observation of the flock on range will soon reveal. A flock queen will not give of her best if a couple of goatlings are milked before her. Her protests will be broadcast to the whole flock, who will then be more concerned with her complaints than with the milking routine.

An earnest silence upon the part of those doing the milking is not necessary. Peace in the milking-shed there should be; but peace and silence are not synonymous. Men and women have not been singing at the milking for the past thousand years or so, merely to exercise their lungs and lose milk. A consistent singer can make the milk drip from the goat's or cow's teats merely by singing the accustomed songs. For a goat distracted by a breach of routine, or by a tender udder, the accustomed song or the regular patter of soothing nonsense is the best aid to let-down available. There is every reason for excluding strangers from the milking shed, but the silent milkmaid is a grim, unhistorical figure with little to recommend her.

The one sound which is quite unforgivable in the milking shed is a sneeze: apart from hygienic considerations, this sound resembles the goat's alarm signal, and invariably (and literally) causes consternation.

The actual process of milking is a knack which is quickly acquired if the basic principles are understood. Anyone who is accustomed to milking cows must realize that they are handicapped by a habit of milking unsuitable for the goat, and pay particular attention to the principles of goat milking.

Grasp the teat lightly in one hand and press the hand gently upwards towards the base of the udder, so filling the teat with milk. Then close the index finger tightly around the neck of the teat, with the hand still pressing gently upwards. This action will trap the milk in the teat. Now close the other fingers in succession tightly around the teat, so forcing the milk down and out. Release the grasp of the teat, and relax the upward pressure. Then repeat the process with the other hand and the other teat.

As the udder empties, the upward pressure of the hand becomes firmer and the initial movement of the milking hand becomes a gentle upward punch. When the flow begins to subside, let go both teats; and, with both hands, gently massage from the top and back of the udder down towards the teat, two or three times. Continue milking until the flow subsides once more; repeat the massaging movement—and so on until no more milk is drawn. The goat should be milked right out to maintain butter-fat content of the milk and to sustain the yield—but 'stripping', as practised on the cow, will distort a goat's udder in a very short time.

Many goatkeepers are bad milkers; a novice may be better advised to follow this advice rather than doubtful example.

In a goat with very short teats it may be impossible, especially for a man, to get more than one finger of the hand around the teat. In this case, on the first upward pressure of the milking hand, trap the neck of the teat between the index finger and the knuckle of the thumb; then extract the milk from the teat by rolling the index finger down the teat while keeping the teat pressed against the thumb.

Men are usually rather handicapped in milking goats for, as often as not, the teat is too short to accommodate all four fingers of a large hand—which means more work, and more fatigue for the other three, and less speed. There is a strong temptation, in the circumstances, especially after the first flush of milk has been drawn, to 'stretch a point' and grasp a portion of the udder above the neck of the teat, so as to bring all four fingers to bear. This procedure is dangerous, for it is quite easy to crush the substance of the udder and force a portion of it down into the teat canal, with disastrous consequences.

It is of no importance whether the goat is milked from the right side or the left, provided she is milked consistently from the same side. A goat with an udder of reasonable shape and capacity cannot be milked from behind without great difficulty and damage to the udder. Scrub goats with low yields, and goats with cleft, pendulous bags, can be so treated without harm or trouble. The advantages of this method will be conspicuous to the intelligent beginner. A restive goat cannot kick over her bucket or put her dirty foot in it when milked from this angle. She can, however, drop dung pellets into the bucket, and even make her water into it if the milker is insufficiently alert. The method, though popular with Mediterranean goatkeepers, is not recommended; but really evil-tempered goats are exceedingly difficult to milk in any other way.

The nervous, fidgety, 'ticklish' goat, usually a first kidder, or one suffering from udder injury, can be dealt with by milking with one hand and gripping the goat about the thigh with the other, so as to encompass the hamstring. Compressing the hamstring will smother a kick, even after it has got under way.

More violent objectors may need to have both hind legs, or all four legs, strapped into position on the milking-stand. Even then, it may be necessary to sling a strap under the goat's belly to stop her lying down. But when this stage of belligerence has been reached, unless the goat has a really painful excuse for her behaviour, it is better to abandon mechanical methods of control and resort to psychological methods (p. 342).

A great deal of trouble at milking time can be avoided if goatlings are accustomed to have their udders handled and massaged long before they are due to kid. The first step in milking a nervous and fidgety goat should always be to stroke and massage the udder until she quiets down: talk to her as you do so, and keep on talking (or singing) as you start milking.

Another great aid to milking without tears or spilt milk is to let the kid pioneer the job for the first four days of lactation, which are the most painful and nerve-racking for the new milker. Most seasoned goatkeepers prefer to put their goats on to a raised platform to be milked; the beginner may prefer to keep the goat on the floor where he can exercise firmer control; but, with the goat standing at a higher level than her milker, there is less disturbance of her coat, and the milk is cleaner; and latterly, when we squat down to goat level, we all tend to creak.

Goats are generally easier to milk than cows; and the worst of them, though tiresome, cannot do any hurt to anyone. The problems created by the difficult milker have been treated in some detail because this is an aspect of reality which goat literature usually tends to gloss over.

GOATS'-MILK RETAILING

Goats'-milk retailers are the main advertisers of the industry; and should never forget the fact, in their own interests and those of their colleagues. Advertisement may be good or bad; a system of accreditation for goats'-milk suppliers, voluntary and under the control of the British Goat Society or local authorities, would probably be of more value to the industry than milk recording.

The law applies no regulations to milk sold direct to the consumer; but if it is sold to a retailer, cartons must be used bearing the name and address of the producer and stating the contents and weight.

Provided that the milk is hygienically produced, the essential point is that it should be distinctively labelled. There is more sales punch in the bottle of still-sweet goats' milk, standing beside the bottle of sour cows' milk of the same age, than there is in a whole-page 'spread' in a national daily newspaper. Clean goats' milk will keep a great deal longer than the best of cows' milk—my personal record, with milk produced under conditions of scullery dairying, was ten muggy July days of a camping holiday.

The seasonal variation in milk supply, which is such a handicap to goat dairying, can be mitigated by stretching the natural breeding season to the limit, by 'running through' (biennial breeding), and by out-of-season breeding; but the variation must remain wider for goats, with their limited breeding season, than for cows who breed all year round. It is some compensation that goats' milk, unlike non-homogenized cows' milk, can be satisfactorily stored in a deep freeze.

Cows' milk will not freeze successfully, because the quick rise of cream causes the formation of large fat globules in the 'reconstituted' liquid. Goats' milk will freeze successfully, provided (a) that it is cooled and frozen immediately after milking; (b) that it is maintained at a steady 0°F (−17·8°C); (c) that it is defrosted slowly. Carelessness at (a) results in fat clusters; carelessness at (b) or (c) results in flaky curd.

Freeze in polythene bags, not more than two-thirds full, to allow for expansion. Date every bag, and allow a cold-store life of up to three months.

There is a higher incidence of off-flavours in goats' milk than in cows' milk; most of them can be prevented by strict attention to the points of dairy hygiene mentioned above. But there is a peculiarly bitter flavour which can occur quite suddenly in the milk of any goat, however scrupulously the dairying is managed. This is thought to be caused by faulty fat metabolism, due to slight anaemia, acetonaemia, or diabetes, and is especially common in newly kidded goats. The cure is to be found in Chapter 10; here it is to be noted that the milk of all goats should be tasted regularly before being offered for sale.

GOATS' CREAM

This is a relatively easy form of produce both to make and to market. If the skim milk can be fed to remunerative stock—calves, pigs, orphan lambs, pedigree pups and kittens, mink, pedigree chicks, etc.—the cream can be sold to compete in price with cows' cream. It will always compete in quality: children prefer goats' cream because it is more digestible; housewives prefer it because it whips to greater bulk. In flavour it is not readily distinguishable from cows' cream; but, unless anatto is added, goats' cream is dead white in colour.

The prime requirements of the cream producer are: (a) a remunerative outlet for skim milk; (b) a strain of goats with high butter-fat milk; (c) a goat diet rich in iodine. For mass production, it is also essential to have a good separator: not all separators are satisfactory for separating goats' milk. Alfa-Laval, Gascoigne, and Lister separators give good results; the farm (over 2-gallon [1·1 litres]) models give better results than the smaller models. A fine setting of the cream screw is necessary in all models, for the fat globules of goats' milk are small, and slow to respond to centrifugal force. If in doubt whether a given separator or screw setting is giving satisfactory cream extraction, run the separator at full speed, and let the milk tap run at half or quarter its capacity: if you get a better cream output per gallon in this way, your cream-screw setting is too wide, or your separator is not readily adaptable to goats' milk. A cream-screw setting or a separator which gives good results at the start of lactation may give poor results later on, for the size of the fat globules in the milk decreases as lactation advances.

The type of cream produced must depend on the taste and nature of the market; the creams which are subjected to much heat treatment before or after separation do not whip satisfactorily, and when they sour they produce foul flavours. The creams subjected to little heat treatment do not help so well; but, if cleanly produced, they sour pleasantly, and for whipping they are superlative.

THIN CREAM
Run the milk through the separator while still warm. If the milk from more than one milking is used, it should be pasteurized when new by holding it to 180°F (82·2°C) for 30 seconds, then cooled quickly and stored in a cool place until required. Before being put through the separator, the milk should be again raised to 180°F and rapidly cooled to 100°F (37·8°C) before being separated. At least part of the milk used for thin cream production should be 'warm from the goat' when put through the separator; otherwise, the cream will not sour pleasantly.

An extraction rate of 1 to 1⅛ pints (568 to 639 ml) from 1 gallon (4·5 litres) of milk will give cream with a fat content comparable to that of most of the cows' cream on the market. In some districts this product is termed 'ream', the term 'cream' being reserved for:

DOUBLE CREAM
This is produced in the same way, but with a cream-screw setting to produce ¾ pint (0·45 litres) per gallon (4·5 litres) of milk.

THICK GRANULATED CREAM
Proceed as for double cream, then immediately cook in a bain-marie or double saucepan for 20 minutes at 180° to 200°F (82·2 to 93·3°C). Cool rapidly, and place in a refrigerator for 12 hours at least. This form keeps extremely well, and is ideal for eating with fruit; when whipped, it forms butter very rapidly, but the butter has poor keeping qualities. Still the best form of cream for long-distance

marketing; it would stand summertime transport from the Hebrides to London.

DEVONSHIRE CREAM

Cool fresh-drawn milk quickly, and leave it to stand in shallow pans in a cool place for 12–24 hours, according to air temperature. Then place the pans on slow heat until the surface cream starts to wrinkle and crack; do not let it boil. Cool again quickly, and skim with a skimmer or saucer. The resultant cream can then be sold for quick consumption, or heated in a bain-marie or double saucepan to 180°F (82·2°C) for 30 seconds and re-cooled, for better keeping qualities.

BUTTER

Butter is not easily marketed at an economic price, owing to the competition of imported cows' butter and the operation of the food subsidies. In large cities where there is a significant foreign population of Greek or Cypriot extraction, and specialized provision stores catering for them, a regular market for goats' butter may be developed—goats' butter is held in special reverence by these people, and traditionally occupies a honoured position on the menu for festal occasions.

But well-produced goats' butter is such a superlative product that, once it has been savoured, the best of cows' butter becomes, in comparison, uninteresting grease. The goatkeeping household will inevitably demand it.

For occasional production, the flavour of the butter can usually be safely left to natural souring organisms, provided there is a high standard of dairy hygiene throughout. But for regular production, it is necessary to control the souring organisms and use a dairy butter starter.

NATURAL-SOUR BUTTER

Separate the milk as for double cream, adding fresh batches of cream daily, or allowing the cream to stand until it smells slightly acid but does not taste sour. Take the temperature of the cream with a dairy thermometer (a cheap implement that saves an immense amount of work in butter making). Bring the temperature of the cream to about 57°F (13·9°C) by standing the can in warm or cold water. If the air temperature is above 60°F (15·6°C), 56°F (13·3°C) is warm enough for the cream; if the air temperature is below 50°F (10°C) bring the cream to 58 or 59°F (14·4 or 15°C). Put the cream in the churn, and churn for about 15–20 minutes—an egg whisk and a bowl make an adequate churn for small quantities. As soon as granules appear in the thickened cream—the noise of churning will change its note at this stage to a 'hollower' sound—add 1 teacup of water for each 1¼ pints (710 ml) of cream. If the air is warm (above 60°F), the water should be cold; if the air is cold, the water should be at about 60°F. Continue churning till the butter granules are the size of peas; then strain off the buttermilk, and wash the butter in successive rinses of cold water until the water runs clear. Then, and only then, gather the butter granules into one lump, and start working the lump in further rinses of cold water until no further buttermilk is exuded. Lastly take out the butter and work it on a board with wooden pats (Scotch hands) which have been soaked in brine, to express all the water. Make up into shapes, and leave to stand 12 hours in a cool dairy.

If preferred, 'salt' dairy salt should be added to the last rinse of the butter granules, and the butter granules left to stand in this for 15–30 minutes.

If difficulty is experienced in getting the butter to break, a not uncommon trouble when the fat globules become very small in late lactation, use the alternative method.

SALTED-CREAM BUTTER
This method is specially designed to ensure the production of well-flavoured butter from small quantities of cream accumulated over a period of a week. The essential factor is that the daily accumulation of cream should be reasonably uniform—say a pint (0·6 litres) a day.

Extract the cream as for natural-sour butter. To the first batch of cream, add 9 oz (255 g) of dairy salt per lb (454 g) of cream, or 11 oz (312 g) per pint (0·6 litres); store in a cool, dust-free place with good ventilation; each day stir in another batch of approximately the same volume as the first, but no more salt. At the end of the week, churn as for natural-sour butter. In cool conditions, the accumulation can continue for a longer period, but the amount of salt used must be adjusted so that the amount of salt added to the first batch is 8 per cent of the weight of the final accumulation of cream. The principle of the system is that, under normal dairy conditions, cream will not sour if the salt content is over 8 per cent.

STARTER-SOUR BUTTER

Separate the milk as for double cream. Place the cream in a double saucepan or bain-marie, and heat to 180 to 200°F (82·2 to 93·3°C) for 20 minutes. Cool rapidly, and place in a refrigerator for 12 hours. If no refrigerator is available, heat to 180°F for 40 minutes and cool rapidly, as low as water temperature will take it. Then warm the cream to 50°F (10°C) and add dairy starter. Cover, and leave to stand until the cream smells acid. If a full-flavoured butter is liked, the cream may be left until it tastes slightly sour (but only when dairy starter is used). Then proceed as in previous method.

If colour is required in either butter or cream, the anatto should preferably be added to the milk before separation: it may be added to the cream before churning.

CHEESE

Cheese is a goat product for which there is the most constant demand, if not always the largest market. From a nutritional view, cheese is superior to any type of meat as a source of protein; and, in this country, it will always be cheaper to produce cheese than its protein equivalent in meat. From a gastronomic standpoint, cheese made in part or whole from goats' milk provides a most effective relief from the dull monotony of factory-made Cheddar and Danish Blues. There is undoubtedly an opening for enterprising cheese makers in this country to produce cheeses of Continental quality, for which upwards of 25 per cent of goats' milk is either essential or desirable.

But the production of any kind of cheese on a commercial scale requires special skill, and a great deal of technical knowledge in so far as the hard cheeses are concerned. The character, texture and flavour of a hard cheese depend upon precise control of temperature and acidity; such control is beyond the capacity of the amateur, and the professional is unlikely to benefit from any advice the present writer can offer.

There remain a few recipes for hard cheese, semi-hard cheese and soft cheeses, which do not require a great deal of skill in the control of acidity and temperature, and which are suitable for cheese making with small quantities of milk. Some samples are given below. The quality of the resulting product will in most cases depend on the type of organisms responsible for 'ripening' the cheese milk. The goatkeeper whose milk normally and regularly sours to a pleasant flavour can risk the use of naturally soured milk as a ripening agent, for occasional cheese production. But anyone seriously intending to produce cheese regularly for the market is advised to cut the risks and improve the uniformity of the product by using a dairy 'starter'.

Management of the starter: the starter is obtainable from dairy supply companies, and will not cost much more than £1·50 per dried packet. It is simply a bacterial culture of suitable ripening organisms. Once activated, it will keep only for a short length of time, but it can be propagated as follows. Simmer fresh milk for 30 minutes, cool to 80°F (26·7°C), add 2 tablespoonfuls of starter per pint (0·6 litres) of milk and keep at 70 to 80°F (21·1 to 26·7°C) till the following day, when the milk will have been ripened to a smooth creamy consistency, and constitutes a new supply of starter. The process may be continued for months at a stretch provided the job is done in clean, dust-free surroundings, and the culture is kept at the correct temperature during the incubating process. Eventually the starter will lose its vigour and flavour, and a new starter will be required.

If starter is not available for use in any of the recipes given below, twice the recommended quantities of buttermilk or naturally soured milk, or ten times the quantities of milk held over from the previous morning, may be used instead of starter—with the risks of failure implicit in the use of uncontrolled ripening organisms.

Rennet: only cheese-making rennet of a reputable brand is suitable. The grocer's junket rennet is quite useless.

Anatto: a little colour improves the appearance of most cheeses. *Cheese anatto* only should be stirred into the milk 10 minutes before renneting.

Salt: use dairy salt only.

The cheese milk: with a few specified exceptions, fresh whole milk is required for all the recipes given. A high standard of dairy hygiene is required to produce good cheese milk. Dirt bacteria cause foul flavours. If there is any doubt about the quality of the milk, it is better to drink it than use it for cheese making. Pasteurization of the milk alters the character of hard cheese, and is responsible for most of the unpleasant and soapy characteristics of the worst factory-made cheese. However, where milk has to be held over from day to day, it is preferable to pasteurize than to risk the milk becoming over-ripe. When pasteurizing for cheese making hold the milk at 145°F (62·8°C)—no higher—for 30 minutes.

The first four recipes given are suitable for goats' milk or for a mixture of cows' milk and goats' milk. The remaining recipes are for 100 per cent goats' milk.

CROFTER CHEESE

A semi-hard cheese of mild flavour maturing in about four weeks and yielding about 18 oz (510 g) per gallon (4·5 litres); requires about 2 gallons (9·1 litres) of milk. This cheese is of better keeping quality and flavour than that produced by the majority of simple cheese-making recipes.

Equipment: a tub or pail, a long bread knife, two 15-in (38 cm) squares of smooth cotton cloth, a perforated steel mould 5¾ in high × 4 in (146 × 102 mm) diameter, 3 weights of 28 lb (12·7 kg) each—or a 6-ft (1·8 m) plank with one end hinged to the wall and a 28-lb weight on the other (this exerts a pressure of 28 lb at the weighted end, 56 lb (25·4 kg) at 3 ft (0·9 m) from the weighted end, and 84 lb (38·1 kg) at 4 ft (1·25 m) from the weighted end, when the plank is kept horizontal).

Procedure: strain and cool 1¼ gallons (5·7 litres) of evening milk. In the morning, warm it to 70°F (21·1°C), and add 4 tablespoonfuls of starter. Strain into it 1¼ gallons of morning milk, adjust the temperature of the mixture to 86°F (30°C) and maintain this temperature. Then, 1¼ hours after adding the starter, stir in 5 drops of anatto; 1½ hours after adding the starter, add 1 teaspoonful of rennet diluted in 6 teaspoonfuls of cold, clean water. Cover and keep warm for 40 minutes. Then test the curd: insert the index finger obliquely into the curd, run the thumbnail along the surface of the curd above the inserted finger, then raise the finger. If the curd is ready for cutting, it will break cleanly and no curd will adhere to the finger; if it is not ready, test again at 2-minute intervals until it is.

When the curd is ready, cut it with the bread knife into vertical slices about ¼ in (6 mm) thick; then cut the slices across to make ¼-in square columns; then gently turn the columns on their sides with the hand or the flat of the knife and slice them up into ¼-in cubes. The job must be thoroughly done and takes time—about 15 minutes. The remaining lumps of curd which you cannot get at any other way should be sliced against the hand. Stir gently for 5–10 minutes.

Now remove 1 pint (0·6 litres) of whey in an enamel or other jug, and place it in hand-hot water, leaving it there until the temperature of the whey in the jug is 105°F (40·6°C). Return the whey to the curd, remove another pint and repeat the process, removing up to 1 quart (1·1 litres) of whey in the later stages, until the temperature of the whole is 94°F (34·4°C). This should take about 45 minutes; a little longer is a lot better than a little shorter. Stir gently while this heating process is going on and, after it is completed, stir for 15 minutes. Then let the curd settle, cover the container and leave for 20 minutes.

Now test the curd for draining. Take a small piece of curd, squeeze it in the hand, and press it briefly but firmly against the surface of a hot, clean poker (or electric iron). Draw the curd gently away from the poker, and if fine threads barely ⅛ in (3 mm) long are formed, the curd may be drained immediately. Strain the whey off into another vessel through a straining cloth.

Bundle the curd in the straining cloth, put a warm pudding basin upside down in the cheese-making bucket or tub, and place the bag of curd on top of it. Cover the bucket. At 15-minute intervals during the next hour, open the bundle of curd, tear the lump of curd into fist-sized pieces and replace them in the bundle, turning cold faces to the inside of the mass. A 7-lb (3·2 kg) weight placed on top of the bundle will assist matting.

At the end of the hour, test the curd against the hot poker again. When the threads so formed reach ½ in (13 mm) in length, the curd is ready for salting. Break it up into walnut-sized pieces, add a good tablespoonful of dairy salt, and stir the salt into the curd for a minute. Bundle up the curd again, leave it for 5 minutes, and give it a final stir before packing into the mould. Do not let the curd get cold.

Warm the mould in tepid water, and press the curd into it by handfuls, pressing just hard enough to cause a little whey to ooze out. Finish by heaping the centre, and placing the lid in position. Apply 28-lb (12·7 kg) pressure and leave for 1 hour. Then turn the cheese in the mould, inserting a fresh, dry smooth cloth as a liner, apply 56-lb (25·4 kg) pressure and leave till next morning. Turn the cheese again the following morning, renew the cloth, and apply 84-lb (34·1 kg) pressure. The same evening or the following morning, the cheese should be bandaged.

Rub the cheese with lard, and apply cheesecloth caps to the two ends. Then sew a wide over-all bandage round the cheese, trapping the cap at either end.

Keep the cheese for 4 weeks at least in warm weather, a little longer in cold weather, in cool room temperature (55° to 60°F) (12·8 to 15·6°C). Turn it daily; if mould starts to form, wipe it away with a cloth soaked in brine.

The times given here should be right for a cheese made with mixed milks; when the proportion of goats' milk is higher than 50 per cent, the times given may be slightly shortened—but the heating of the curd must not be speeded, especially in the earlier stages.

Beginners at cheese making will find a clockwork 'pinging' timer a great aid in timing.

(This recipe is the most ambitious given here, and is the most suitable for production for a regular market.)

LITTLE DUTCH TYPE

A softer and more open-testured cheese than the Crofter, with less flavour. The quality of this cheese is even more dependent on the quality of the cheese milk.

Equipment: as for the Crofter cheese.

Procedure: prepare the cheese milk as for the Crofter, but raise the temperature of the mixed milks to 90°F (32·2°C) before renneting. Rennet the milk 30 minutes after adding the starter, using $1\frac{1}{4}$ teaspoonfuls to 1 wineglassful of cold water. Otherwise proceed as for the Crofter; but the curd will require a little longer before it is ready to be cut. In cooking the curd, raise the temperature to 105°F (40·6°C) in 1 hour; the whey should not be heated above 100°F (37·8°C) for the first 30 minutes, nor above 110° to 115°F (43·3 to 46·1°C) for the second 30 minutes. Once the temperature of 105°F is reached, stir the curd for 15 minutes, let it settle, and strain off the whey immediately. As we do not want to bundle the curd in this case, it is better to bale off some of the whey and decant the rest. Immediately add 2 heaped dessertspoonfuls and 1 level dessert-spoonful of dairy salt and a pinch of saltpetre; stir in as for the Crofter. The cheese mould should be at the same temperature as the curd, which is packed into the mould as quickly as may be after the mould is removed from the whey. Pressing, turning and bandaging is carried out as for the Crofter.

Typical 'little Dutch' moulds can be used for this cheese and, provided a good starter is used, the cheese milk can be pasteurized without detracting from the character of the cheese in this case. Unless the cheese milk conforms to the highest standards, it is safer to pasteurize.

WENSLEYDALE CHEESE (*Lady Redesdale's recipe*)

Use 2 gallons (9·1 litres) of milk for 2 lb (0·9 kg) of cheese. Heat milk to 82°F (27·8°C). Add 1 teaspoonful of rennet mixed with 5 teaspoonfuls cold water, and stir for $\frac{1}{2}$ minute. Leave 1 hour.

Cut the curds into $\frac{1}{2}$-in (13 mm) cubes, as for Crofter cheese. Stir gently for 5 minutes with the hand.

Raise the temperature of the curds to 86° to 90°F (30 to 30·2°C) in the course of 20–30 minutes, by standing the curd bucket in a larger vessel of hot water and stirring constantly. When the cubes of curd are firm and 'shotty' leave to stand for 10 minutes. Drain off the whey, put the curds in a cloth, and squeeze out the whey. Lay a 10-lb (4·5 kg) weight on the bundle of curd, and leave for 10 minutes. Break the curd and turn the cold faces inwards, bundle up, and replace weight for a further 10 minutes; repeat this last process. Then break up the curds by hand, and salt to taste. Leaving the curds in the cloth, place cloth and curds in a mould, fold the top of the cloth flat, and apply a 10-lb weight. Turn daily for 3 days, replacing the weight each time. After 3 days, remove the cheese from the mould, and place it on a shelf in a cool, airy place, turning it daily. It will be ripe in 3–4 weeks.

BRIE-TYPE (*Recipe from a long-established Anglo-Nubian goat dairy in Eire*)

Bring 2 gallons (9·1 litres) of perfectly fresh, clean milk, preferably warm from the goat, to 83°F (28·3°C). Mix 2 ml cheese rennet in 10 ml of water. Stir into the milk. Leave to stand for 2 hours.

Ladle the curd into hoops, 10 in (25 cm) in diameter and 3 in (8 cm) high, standing on a straw mat on a draining-board. Allow to drain for 24 hours, or until the curd is firm enough to hold its shape without the hoop; sprinkle salt on the upper surface of the cheese, and leave for another 24 hours. Turn the cheese on to a fresh straw mat, and rub salt into the second side. In a few hours turn again and leave to ripen.

Turn the cheese each day in a dry well-ventilated atmosphere. In about 8 days, moulds begin to grow on the surface; the cheese should then be transferred to a dark cellar with little ventilation and with a temperature around 55°F (12·8°C). The ripening is brought about by the moulds, which grow on the surface, diffusing enzymes into the cheese. If conditions are not right for the moulds, the cheese will spoil; mould growth is the key to the operation. The cheese should ripen in 2–4 weeks, by which time it should have a semi-liquid or waxy centre.

PONT L'EVÊQUE-TYPE

This resembles the distinctive Normandy cheese which is made from cows' milk. Yield: 2 cheeses about 5 in (13 cm) square and 2 in (5 cm) deep, weight 15 oz (425 g) from 1¾ gallons (7·9 litres) of milk.

Equipment: 2½-gallon (11·4 litres) pail for the cheese milk, a sharp-edged skimmer. A large tub or dairy sink, in which the pail can be immersed (not necessary in warm weather); a bread knife for cutting the curd; 2 yd of 36-in (1·8 m of 0·9 m) cheesecloth; straw or bamboo mats, 1½ in × 8 in (38 mm × 20 cm); draining boards 14 in × 8 in (35 × 20 cm)—preferably 4 mats and 4 draining boards for each pair of cheeses, but two will do. Two steel moulds, square with rounded corners, 4⅞ in square × 2¼ in (124 × 57 mm) deep; a draining rack about 30 in × 12 in (76 × 30 cm) with ⅜ in (9 mm) spars set about ½ in (13 mm) apart. (All standard cheese-making equipment obtainable from dairy suppliers.)

Procedure: strain the morning's milk—1¾ gallons (7·9 litres)—into the cheese pail; adjust the temperature to 92°F (33·3°C) and place the pail in the tub or sink half-filled with water at 94°F (34·4°C). Or work in a warm draughtless room during warm weather. Add 2 drops of starter and 15 drops of anatto, stirring in each in turn. Dilute ½ teaspoonful rennet in ½ wineglassful of cold water, and add this to the cheese milk; stir for 5 minutes. Cover the pail with a cloth. Leave for 45 minutes; but in cold weather, remove a little of the water in which the pail is standing every 15 minutes, and replace with water at about 94°F. The temperature of the water may be allowed to drop, but no quicker than it would drop on a good spring day.

After the 45 minutes, test the curd for cutting as in the Crofter cheese; it may require up to 1 hour before it is ready; if it takes any longer, the cheese milk has not been kept sufficiently warm. When the curd is ready, cut it into 1-in (2·5 cm) square columns, and then cut diagonally through the squares, once only, to give triangular columns with 1-in sides. Leave for 10 minutes.

Then, with a skimmer, ladle the curds in ¾-in (19 mm) slices into a previously scalded straining cloth spread over the sparred rack. Fold the ends of the cloth over the curd, cover with another dry cloth, and leave for 30 minutes.

Open the cloth, and quickly cut the curd into 3-in (8 cm) squares. Fold back the cloth and re-cover with the dry cloth. Leave for another 30 minutes.

Take one corner of the folded cloth and bind it tightly round the other three, bundling the curd slightly in the process. Re-cover with the dry cloth. Leave for 15 minutes. Tighten the bundle and turn it upside down, with the knot underneath. Leave for another 15 minutes.

Place the 2 moulds together on a scalded straw mat, and the straw mat on a draining board. Then open up the curd, breaking it into strips and small pieces as you transfer it into the moulds. The curd lying against the side of the mould should be firmly pressed against the side; as each layer of curd is laid, sprinkle it with dairy salt, using 1 heaped tablespoonful of salt to each cheese.

When the moulds are filled, cover them with a scalded straw mat, with a draining board on top of that. Now comes the beginner's anxious moment. Grip the two boards between the fingers of both hands, and turn them over so that the bottom of the cheese is now uppermost. Repeat this process at 10-minute intervals for the next hour, and once more before you pack up for the night.

Leave the cheeses in a cool, airy room, turning them in the same way each day; it is preferable to change the top mat and draining board for a fresh and scalded pair if the weather is warm and muggy. After 4 days from the making of the cheese, remove the moulds. Smooth the sides of the cheese with a knife to fill in any crevices, lay the cheeses on a dry muslin and turn them each day until the surface is quite dry—about a fortnight if conditions are favourable.

Now wrap each cheese in greaseproof paper and wrap each pair of cheeses, or preferably each 4 cheeses if you have them, in a cheese cloth, scalded and dried. Place the bundles so that the cheeses are lying on their narrow sides, on a latticed shelf or baking tray in a cool, airy room. Every day, turn the bundles round so that the cheeses lie on another part of their sides; every alternate day open the cloth bundles and turn the cheeses so that the side of the cheese facing inwards now faces outwards. If there are 4 cheeses in the bundle, move the inside ones to the outside.

Ten days after the cheeses are wrapped, they should be unwrapped completely, and any mould growth on them wiped off with a cloth soaked in boiled and cooled brine. A cheese with much mould growth on it should be dipped in the brine solution and allowed to dry before being re-wrapped in clean paper. If, while turning, any mould is noticed on a cheese at an earlier date, it should be treated in the same way.

The cheese should be ripe 2–4 weeks after wrapping, and is best eaten within the following fortnight—hence the reason why the cheese is not widely imported.

The Mont d'Or

France's most popular 100-per-cent goats'-milk cheese is made in a very similar way, but by people who apparently attach no particular value to a 50-hour week. I have been unable to obtain a detailed recipe, but the following may provide early risers with a basis for experiment.

To 2 gallons (9·1 litres) of fresh clean milk at 82°F (27·8°C) is added 1 drop of starter and 4 drops of rennet diluted in 2 teaspoonfuls of water—each being thoroughly and separately stirred in. The cheese milk is kept in a warm place, and the curd is ready for cutting in about 8 hours, or a trifle longer. The curd is then cut as for the Pont L'Evêque, and is strained in a 9-in (23 cm) long, conical strainer for 4 hours, no attempt being made to conserve the heat of the curd. Thereafter, the curd is packed into circular moulds a little over 4 in (10 cm) high and about 3 in (8 cm) in diameter, with 2 heaped tablespoonfuls of salt sprinkled between the layers. A scalded condensed-milk tin with the rims removed would seem to provide an experimental mould. The moulds are placed on and covered by the same arrangement of straw mats and draining board as the Pont l'Evêque, and the cheeses are frequently turned in the same way. After 3–4 days, when the cheese can hold its shape without the aid of the mould, the mould is removed, and the cheese is rubbed with salt and dried for a further few days on clean muslin laid across a slatted rack or baking tray. When sufficiently firm, each cheese is completely enveloped in a cheese bandage or wicker basket, and hung up to dry in a cool, airy room. The cheese ripens about 10–15 days after leaving the mould in the summer of central France, probably a little longer in our climate.

With a mould of this height, a few elastic bands strategically stretched between the draining boards would appear advisable.

ROSS-SHIRE CROWDY BUTTER

This is not, strictly speaking, a cheese as we know it; but it is a modernized version of an eighteenth-century Highland recipe for a curd-and-butter mixture, based on the use of goats'-milk curd and cows'-milk butter. It produces a mellow, highly flavoured product somewhat akin to some of the Flemish cheeses.

To 1 gallon (4·5 litres) of fresh milk at 90°F (32·2°C) add ½ pint (0·3 litres) of starter, stir thoroughly, and leave in a warm place for 6–12 hours, when the whole will have become a smooth, creamy curd. Heat to 110°F (43·3°C) by standing it in hot water and stirring; maintain this temperature for 30 minutes–1 hour, stirring continuously until the curd hardens sufficiently to drain off the whey. When the whey has been drained, the curd is washed with sufficient cold water to reduce the temperature to 70°F (21·1°C) and strained through a cloth. The cloth containing the curd has its four corners tied together, and is hung up to drip for 1–2 hours, when the curd should be of firm pasty consistency.

The curd is then thoroughly mixed with 1 heaped tablespoonful of salt and ¾ lb (340 g) of butter. A scalded wooden box 7 in long, 1½ in wide and 4 in deep (178 × 38 × 102 mm) is lined with two strips of cheesecloth that have been scalded in boiling brine, the one strip being 1½ in wide and 23 in (58 cm) long, the other 7 in wide and 12 in long (18 × 30 cm). The curd and butter mixture is rammed into the box tightly, with particular attention to filling the bottom corners, and the loose ends of the cheesecloth are turned over to cover the top of the mixture. Any space remaining at the top of the box is packed with salt, a lid is fixed and tied firmly down, and the box is stored in a cool airy place for 4–6 weeks: 4 weeks is sufficient for the first trial.

If this recipe is cut short after the salt is added to the curd, without the butter, you have the recipe for crowdy or cottage cheese, which must be eaten within a few days of making. It is, incidentally, the better for having the butter worked in, even if it is to be eaten fresh. Skim milk can be, and frequently was, used for this recipe.

NORWEGIAN WHEY CHEESE

This is a useful method of using up whey. Evaporate on low heat, stirring often until creamy, and thereafter continuously until pasty in consistency. Spoon into greased bowls to cool; tip out. It is greyish-brown, with a concentrated sweet-salt-sour flavour, and very nutritious; the market is small, but enthusiastic.

SMALLHOLDER CHEESE

To 2 pints (1·1 litres) of fresh warm milk add ½ teaspoonful of rennet and stir well. Let stand for 12 hours, then cut the curd into pieces of uniform walnut size. Transfer the cut curd to cheese cloth and hang it up to drain for 24 hours. Then add a pinch of salt to the curd and place it in a mould stood on a small straw mat. Place another mat on top and slightly weight down.

After one day turn the mould right over, use clean mats and weight from the other side to compress the curd.

After a further day the cheese is ready for use.

LACTIC ACID CURDS

Use whole, skim or separated milk. Heat to boiling point and make acid in one of the following ways:

(a) 2 tablespoonfuls of vinegar to every 3 pints (1·7 litres) of milk.

(b) 2 tablespoonfuls of lemon juice to every 3 pints of milk.

(c) ½ teaspoonful of tartaric acid dissolved in a cupful of hot water and added to 3 pints of milk.

Stir in the acidifying agent and the milk protein will curdle immediately and the whey separate out. Strain through a cloth or fine nylon sieve.

This curd will have little flavour and can be used in sweet or savoury dishes, such as the basis for Yorkshire curd/cheese cakes; or add chopped herbs (chives, lemon balm, etc.) and use as a spread.

It can also be pressed under a 4-lb (1·8 kg) weight, then cut into cubes and used in curries.

Mrs Leueen Hill of Redruth, Cornwall, kindly supplied these last two recipes.

YOGHOURT

First made in the Balkans, and with goats' milk, which results in a slightly less acid produce. The essential element is rapid conversion into preservative lactic acid of the lactose (milk sugar) by a pure culture of bacteria which leaves all other constituents unaffected. Whereas other preservative treatments impair flavour and nutritive qualities, this process preserves and enhances protein content, and leaves fat content and vitamin potency intact.

To make yoghourt: bring 1 pint (0·6 litres) milk to the boil and simmer for 1 minute. Leave milk to cool until it reaches 98·4° to 100°F (37 to 40°C). Put 1 generous dessertspoon of live yoghourt into a bowl and slowly add the cooled milk, stirring thoroughly so that the two are well mixed. Transfer the contents to a wide-mouthed vacuum flask, screw on the lid and leave the flask to stand for about 8 hours. This is an extremely reliable method, but in the absence of a suitable flask the bowl can be covered tightly with a plastic bag, wrapped in newspaper or blanket or towelling and put in a warm, draught-free place (an airing cupboard or the back of a solid fuel stove, for instance) and again left for 8 hours, or a little longer if necessary.

Milk must be initially clean and kept free of contamination by other bacteria.

Various strains of yoghourt bacteria are available from health-food shops and specialist suppliers; each imparts a slightly different flavour, and demands a slightly different treatment, which will be specified. For goats' milk, the Bulgarian strain is best.

True yoghourt is an almost pure culture of the specific bacteria, and can be used as a 'starter' for further batches. Pseudo-yoghourt, acidified milk totally without live yoghourt bacteria, is easy to produce; easiest of all if lactic acid from the chemist is added directly to the fresh milk seasoned to taste. Its flavour is indifferent; and if you try to incubate the sterile pseudo-yoghourt with fresh milk, it just curdles deceptively.

Chapter Twelve

The Universal Foster Mother

To the supreme honour of a place in the heavens, among the signs of the Zodiac, the Greeks elevated three of their domesticated animals: the Bull who drew their ploughs; the fleecy Ram who clothed them; and Capricorn the Goat. The name of the goat who earned this honour for her species was Almalactea 'Foster milk'. Her constellation still brightens the twentieth-century sky.

While relatively few of the newly born of other species can be satisfactorily reared on cows' milk, however modified, there is probably not a single species among the larger land mammals the young of which will not thrive on suitably adjusted goats' milk.

The purpose of the present chapter is to consider the adaptation of this highly digestible food to the peculiar needs of the various species of farm livestock, and to suggest how Almalactea can turn an honest penny for the twentieth-century farmer.

It is generally accepted that the composition of the milk of each species is ideally adapted to the growth pattern of the young of the species, and that any other milk composition will represent a departure from the ideal.

In other words, goats' milk will be entirely suitable for youngsters of other species which grow at approximately the same rate as the kid: for those that grow faster, it will prove too weak; for those that grow more slowly, it will prove too strong.

It is perfectly true that members of the medical profession are, for the most part, convinced that cows' milk, which is designed to suit the fast growth rate of the calf, is perfectly suitable for feeding the slow-growing human infant. It is perhaps sufficient to point out that farmers are concerned with the economic life history of their stock, while doctors are in practice primarily concerned with resolving immediate problems, and are seldom able to test the long-term effects of their recommendations. There is a massive collection of scientific evidence to show that all young farm stock, if fed a diet too rich in digestible protein, are liable to mineral-deficiency disease sooner or later.

No one has any doubt that all youngsters must be fed a diet sufficiently rich to maintain normal rate, if they are to be healthily reared.

Table 10 shows the comparative composition of the milk of the various species of domesticated animals, including women (!).

Table 10: Comparative composition of the milk of various species

Species	Water (%)	Fat (%)	Sugar (%)	Casein (%)	Other protein (%)	Ash (%)
Goat	86·2	4·5	4·08	2·47	0·43	0·79
Cow	87·3	3·67	4·78	2·86	0·56	0·73
Sheep	79·46	8·63	4·28	5·23	1·45	0·97
Mare	89·8	1·17	6·89	1·27	0·75	0·30
Donkey	89·88	1·5	6·09	0·73	1·31	0·49
Dog	75·44	9·57	3·09	6·10	5·05	0·73
Cat	81·63	4·49	4·79	3·72	3·30	0·58
Pig	83·23	4·5	4·2	7·3		0·77
Woman	87·4	3·0	6·5	0·04	0·7	0·25

Table 11 shows the adaptation of goats' milk which is required to produce a milk as nearly as possible ideal for the rearing of various classes of farm stock. It will be seen that this adaptation is quite satisfactory for calves, lambs and foals; it is not so good for puppies, but valuable orphan pups can almost always be reared on this mixture, which gives them insufficient protein, to the stage at which they can take enough of the milk to provide all the protein they need. For rearing kittens from birth goats' milk is the best substitute for cats' milk available, but will only rear the stronger members of a strong litter. Few people will be unduly concerned at this lapse from perfection.

Table 11: Adjustment of goats' milk for the feeding of the newly born of farm animals

Species	First ten days	Thereafter
Calf	Whole milk to appetite in four feeds per day.	Whole milk up to 1 gallon (4·5 litres) per day in two to four feeds.
Lamb	Whole milk with 1 oz (28 g) of thin cream added to each ½ pint (0·3 litres) to appetite in four feeds per day.	Whole milk up to 2 pints (1·1 litres) per day for hill lambs, and up to 4 pints (2·3 litres) per day for other breeds, in four feeds per day.
Foals (including donkey foals)	Half-and-half milk and water with 3 level tablespoons lactose (sugar of milk) to each pint of *milk*, or 2½ level tablespoons sugar to each pint of *milk*, and 1 teaspoon lime-water to each pint of *mixture*, in four feeds of up to 1 pint (0·6 litres).	The same, feed to appetite in four feeds a day. At 2 months the proportion of milk can be gradually increased to 75 per cent of mixture.
Pups	1 teaspoon of thin cream in a tablespoon of whole milk at eight feeds per day; feed to appetite.	The same at six feeds till 3 weeks. Then whole milk to appetite in four to six feeds.
Kittens	Whole milk in six to eight feeds per day.	The same, reducing number of feeds.
Piglets	1 dessertspoon of cream in 1 teacup of milk. 2–4 oz (57–113 g) at six feeds per day. Or 2 dessertspoons Glucodin instead of cream.	Whole milk with 1 tablespoon of cream per pint (0·6 litres) to appetite at four feeds. Trough feed at 3 weeks. Replace Glucodin with brown sugar.

Notes on table:

(1) Colostrum from a newly kidded goat may be given to the new-born calf and the new-born lamb, but not to any other new-born. One feed is sufficient.

(2) All feeds must be fed at blood heat and at regular intervals. But the interval before the last feed at night may be longer than the others. Feeding bottles must be sterilized after each feed.

(3) *Calves*. Give 1 dessertspoon of olive oil or castor oil in the first feed if the calf cannot receive colostrum.

(4) *Lambs*. Give 1 teaspoon of olive or castor oil in first feed if the lamb cannot obtain colostrum. Feed milk in a polythene baby's bottle fitted with lamb teat, or in a nylon 'fre-flo' bottle.

(5) *Foals*. Give a first feed of 1 dessertspoonful of castor oil in *whole* milk if the foal cannot receive colostrum. Feed from a wine bottle fitted with calf teat.

(6) *Pups*. Add 2 drops of cod-liver oil to the first two feeds. Feed with a glass dropper or, in emergency, with a fountain pen. Add 'Sister Laura's Food' for weaklings and miniatures.

(7) *Kittens*. Feed with a glass dropper or fountain pen.

(8) *Piglets*. Add 1 teaspoon of cod-liver oil to first two feeds if the piglet has not had colostrum. Feed with a polythene baby's bottle and baby's or soft lamb teat.

The details given above should suffice to tide the farmer and his orphan over a crisis. But for the goatkeeper who wishes to develop the commercial possibilities of 'foster milk', other considerations are of importance.

The best place for a goats'-milk orphanage is on an unspecialized smallholding. In any reasonably densely populated farming district, there is a reliable if seasonal supply of orphan lambs, orphan piglets and piglets surplus to the milking capacity of the sow, with an occasional orphan foal to add interest. A word to the veterinary surgeon, and a little market-day advertising, will canalize the supply in the required direction. Orphan lambs can be reared on cows' milk, though they seldom do nearly as well as they do on goats' milk. Consequently, orphan lambs command a fair price. Piglets reared on cows' milk from birth never pay their way, and usually die. So orphan and surplus piglets are cheap, and provide the staple throughput of the goats'-milk orphanage.

The first rule of the orphanage should be: fresh orphans only. There is little hope of making a profit out of any new-born creature that has already had its digestive system and vitality undermined by a diet of cows' milk or cows' milk and water, or by simple starvation. This is especially true of piglets, who are not worth having unless they are fresh from the sow.

Piglets will take to the bottle easily. The weaker of them may benefit from the addition of a little glucose to the feeding recipe given in Table 11; but this should not be continued for more than two or three days. The piglet's main need during the early stages is for warmth. For a regular orphanage business, an electric- or paraffin-operated infra-red heater is worth while. For casual business, a straw bed of good depth, with a few well wrapped hot-water bottles under the surface, will suffice.

Frequent feeding of the piglets during the first three weeks is essential. Six feeds at, say, 7, 10 a.m., 1, 4, 7 and 10-30 p.m. will suffice without greatly disrupting the peace of the farmer. After three weeks, four feeds a day will be sufficient; and the milk can be fed in a trough, with a little taste of meal and some milk-soaked bread added as soon as they have become accustomed to the change. The meal ration is increased as the piglets grow; and, by five weeks old, they will be able to maintain progress if the whole milk is replaced by skim.

From a few days old, the piglets should have room for exercise; if weather and circumstances permit, they will do best with a grass run; otherwise they must have some turfs or soil to supply the iron they need. If there is a good supply of piglets, the goatkeeper is best advised to sell at eight weeks old, when the piglets will weigh about 40 lb (18 kg). Sold at this stage they should return a good price in proportion to the milk, etc. that has been fed to them. If kept any longer, the return per gallon will drop substantially and a cream market would have to be found to make the enterprise profitable.

These comments are not intended to describe methods of pig rearing—for which the writer is not qualified—but to clarify the main points of contact between goatkeeper and piglet rearing.

Orphan lambs should always be reared on the bottle, and never suckled on the goat. This is in the best interests of both parties. Ewes have smaller and tougher teats than goats, and a lamb is likely to break the skin of the goat's teat with its teeth, and give rise to black garget infection. The goat's grazing habits are very different from those of the sheep, and a goat suckling a lamb will lead it to pastures unsuitable to the lamb, and teach grazing habits inimical to the lamb's digestion.

Calves will not feature in the goat orphanage; there is plenty of cows' milk. But calf rearing can provide a basis for profitable goatkeeping. We are not concerned here with the farmer who keeps goats and uses their milk to rear his replacement heifers, while drawing the subsidized price for the cows' milk the calves would otherwise have consumed. But a goatkeeper may specialize in calf rearing.

The goatkeeper may reasonably expect to rear a slightly better calf than anyone else for two good reasons; the high digestibility of

Plate 17. R77 Osory Snow Goose, owned by Gary and Chris Jameson, Mytaba Stud, Leppington, New South Wales, Australia. (D. Mytaba Sulpaedes.) Holder of Australian and world records for 1st Lactation and Senior Doe. 7267 lbs on 1st lactation, 7713 lbs on 2nd lactation.

goats' milk will minimize setbacks from digestive troubles, and will practically exclude the normal liability to white scour; the fact that goats and calves share but few internal parasites will afford both parties a measure of protection when sharing a pasture—the calf will eat the larvae of the worms that infest goats, and the goats will eat the larvae of the worms that infest calves, with no ill consequences to either.

Stock rearing of this kind is one of the least profitable forms of farming in these days. If calf rearing is to be really attractive to the goatkeeper, he must be able to rear a calf which cannot be matched by any other method of rearing, and command the premium his supremacy deserves. There is only one way in which this can be done: for a goatkeeper with a herd of Anglo-Nubians or Anglo-Nubian crosses (that is, a herd producing milk with an average butter fat of over 5 per cent), to rear pedigree heifers of a breed giving low-butter-fat milk. A Friesian heifer reared on Anglo-Nubian goat's milk is a better reared beast than any Friesian heifer has a natural right to be. If you put a Friesian calf on a Jersey cow's milk, it will get the butter fat; but it will also get indigestion. Only the goat can do the job better than Nature.

This is not the job for the novice, nor even for the specialist goatkeeper without other experience. The calves worth rearing are worth three figures at birth. But here is an opportunity for a first-class all-round stockman with a good knowledge of goats, and a special knowledge of bringing out cattle, to do an independent job dear to his heart and good for his pocket.

Calves will literally, and properly, go down on their knees for a drink of goats' milk, and may be suckled on a goat with adequate teats. The system has been practised without mishap on a small scale: whether the saving of labour and additional protection from scour provided by direct suckling is worth the risk of injury to the goat's udder, only time and experience can tell.

How the stars of Capricorn will twinkle when the champion dairy cow is reared on goats' milk!

Chapter Thirteen

Kid Meat

The meat of adult goats is not particularly good eating, and is generally only consumed *faute de mieux* in most countries; in Britain, it is held in particular aversion. Earlier editions of this book contained information about it which has now been left out, as being a lost cause. One worthwhile recipe for goat ham is retained.

Kid meat, on the other hand, can be a prime meat, and is an entirely different proposition from any other kind of goat meat. The male kids arrive whether we want them or not, and for four days at least the goat produces milk which is of little use for anything except kid-rearing. So we have a 10-lb (4·5 kg) kid free.

From a culinary point of view, kid is rather more versatile than, and almost as well flavoured as, lamb; for the modern taste for lean meat, kid meat may well prove superior. The flavour of meat depends to a great degree on the condition of the animal from which it is derived; it is not possible to bring lambs to the condition in which their meat develops its optimum of flavour without at the same time introducing more fat into the meat than the modern housewife cares to see. But a perfectly conditioned kid will present a lean chop.

It is worth pointing out that the price of meat in the butcher's shop is today approximately twice the price that the farmer receives per lb dressed carcase weight. So kid meat is worth twice as much if you eat it at home as if you sell it to the butcher.

At present in France, the National Institute for Agronomic Research (INRA) is investigating the increased breeding of goats, of which there are now about one million head in France. Goat production until now has been mainly for the production of cheese, but now consumers are said to be interested in kid meat. Previously heavy kids were considered a by-product of the dairy industry. Suckling kids used to drink their mother's milk for about two weeks before being slaughtered, while weighing just 13 or 15 lb (5·9 to 6·8 kg). Now breeders tend to feed kids with milk products and to slaughter them aged between 3 and 5 weeks, weighing between 18 and 26 lb (8·1 and 11·8 kg) or sometimes even more. Researchers have developed feeding techniques for producing heavier animals. At the same time they have tried to rationalize techniques in order to reduce manpower costs.

Kid-meat production brings one new consideration into goat-keeping practice: the necessity to castrate the male kids at birth. The effect of castration on the growth rate of male kids is far greater than is the case with ram lambs and bull calves. The male kid may be sexually mature at three months; and, at that age, the effect of castration at birth may result in doubling the efficiency with which the kid converts milk into meat. The meat is also of better quality.

By far the simplest and safest method of castration is by the 'Elastrator', a device for applying a specially designed rubber band which cuts off the supply of blood to the scrotum. It should be used within a day or two of birth if possible, and the sooner the better.

From a culinary point of view, kid meat is of three types. Month-old kid, which is a white and rather glutinous meat, is like veal, or chicken in the *poussin* stage. It can be used in most veal and *poussin* recipes, but is best in the more highly flavoured ones. Three- to four-month kid can be treated as spring lamb, which is a singularly happy thought in days when there are very few spring lambs to be treated; the better restaurants and hotels can be very grateful for this type of meat. Six- to nine-month kid can be larded and treated like lamb; but it is at its best when marinated, for then it becomes as well flavoured as prime venison and as tender as prime lamb, and can be used in recipes designed for either meat.

Some recommended recipes are given below.

ROAST KID (*Central France*)
A leg of kid of three to four months old, weighing 3 to 4 lb (1·4 to 1·8 kg) is pierced with the point of a carving knife in eight to twelve places where the flesh is thickest. Into half of the gashes place a piece of peeled garlic the size of a split pea; into the rest of the gashes force ¼-in (6 mm) strips of bacon fat or salt pork.

Place the kid in a roasting tin with ½ pint (0·3 litres) water and 1 teaspoonful of salt; break 1 oz (28 g) of butter into small pieces, and dot them over the top of the joint.

Roast in the oven at 340°F (175°C, gas mark 3½), allowing 20 minutes per lb (454 g).

Remove the joint into a heated dish and keep hot. Scrape the bottom of the roasting tin, add a little water if necessary, strain, and serve this gravy with the meat. Garnish with watercress.

ROAST KID (*French Alps*)

Chop a 1-in (2·5 cm) long sprig of tansy or thyme, a handful of parsley and two cloves of garlic; crush 1 teaspoonful of dill or caraway seeds; mix the chopped and crushed herbs with 2 tablespoonfuls of wine vinegar, 1 teaspoonful of salt and 1 teacupful of olive oil.

Lay a fresh leg of kid of six to nine months old on a dish. Pour the above mixture over it; baste it with this marinade frequently, and turn it daily for 3–4 days, keeping it in a cool airy place.

Pour 1 teacupful of the marinade into a roasting pan. Place the joint in the pan and roast in the oven at 300°F (150°C, gas mark 2) allowing 30–32 minutes per lb (454 g).

Ten minutes before the meat is done, melt 2 oz (57 g) of butter in a saucepan, add a small clove of garlic well chopped; let it simmer 5 minutes; stir in 1 tablespoonful of flour; cook and stir till the flour is golden brown; add ½ pint (0·3 litres) of stock or hot water, and stir vigorously till the sauce thickens; withdraw from the heat, and add 1 tablespoonful wine vinegar, 1 teaspoonful dark honey, 1 tablespoonful concentrated tomato purée—or 2 tablespoonfuls fresh tomato purée—and continue heating till the sauce starts to boil.

Serve this sauce with the roast kid; do not serve the gravy from the roasting dish.

This is kid at its best.

GRILLED SHOULDER OF KID (*an old Highland recipe*)

Detach the shank from the blade of a shoulder of kid of three to six months old—so that the joint is shaped like an axe-head. Lay aside the shank. Rub the joint over liberally with butter.

Place on a grilling rack 3 to 4 in (8 to 10 cm) under the grill which should be moderately hot; turn the joint frequently and baste occasionally; test for readiness by pricking the thicker portions with a fine needle; the exudation should be pale pink.

Serve with rowan or redcurrant jelly.

PEKIN KID (*adapted from a Chinese recipe: suitable for meat from any age of kid*)

Cut the kid meat into small pieces, dip in a thin flour, water and salt batter, and deep-fry till golden brown. Lay aside in a warm place to drain.

Melt 2 oz (57 g) of butter in a large saucepan, add $\frac{1}{2}$ lb (227 g) of carrots cut in fine strips, and a dozen sticks of celery or seakale beet cut in short lengths; cover and cook slowly, tossing and stirring occasionally. Add water only if and when necessary. Cook for 20 minutes.

In another saucepan, heat 3 tablespoonfuls of olive oil until smoking hot, stir in 2 tablespoonfuls of flour, and cook till golden brown; add $\frac{3}{4}$ pint (426 ml) of stock, and stir vigorously until the sauce thickens. Withdraw from the heat and stir in 1 teaspoonful of golden syrup, 1 teaspoonful of French mustard, 1 tablespoonful of mango chutney and a little pepper. Pour the sauce over the vegetables, add $\frac{1}{4}$ lb (110 g) mushrooms and $\frac{1}{2}$ teacupful of chopped chives, and then the fried meat which was laid aside. Cook for 5–10 minutes. Add 1 lb (454 g) of finely shredded cabbage, cover and keep barely simmering for 20 minutes more. Serve with rice.

KID AND GREEN PEAS (*Normandy*)

From a kid about one month old cut small pieces of meat, salt and flour them and brown them in butter in a large saucepan. Add 1 pint (0·6 litres) water per lb (454 g) of meat, and stir till blended. Then add 4 spring onions, one 6-in (15 cm) sprig of parsley, and 1 lb of fresh green peas per lb of meat. Simmer gently for 1 hour.

(If tinned or frozen peas are used, they should be added only 15 minutes before serving.)

KID IN CREAM (*Sweden*)

From a kid of about one month old, cut slices of meat or small joints for serving. Sprinkle all over with salt and pepper and dredge, very lightly with flour. Place in a thick saucepan with $\frac{1}{2}$ cup of thick cream per lb (454 g) of meat and bone. Cook until the meat is browned, turning frequently, and adding more cream as necessary. Cover, and cook gently until the meat is tender.

Remove the meat into a heated dish, and keep hot. Add sufficient flour to the fat remaining in the pan to make a smooth cream. Cook for 3 minutes, Then add equal parts cream and kid stock (or chicken stock), allowing $\frac{1}{2}$ cup of each for every 3 tablespoonfuls of flour added. Stir till the sauce boils. Add chopped parsley, and pour over the meat.

GOAT HAM (*a traditional Highland recipe*)
The hind leg of a twelve- to eighteen-month old castrate (or sterile goatling) is trimmed to shape and rubbed with the following mixture: 1 oz (28 g) saltpetre, 4 oz (110 g) brown sugar, 1 lb (454 g) preserving salt, 1 oz (28 g) white pepper, $\frac{1}{4}$ oz (7 g) of cloves, 1 grated nutmeg, and $\frac{1}{2}$ oz (14 g) coriander seeds. Rub the mixture into every crease and crevice in the flesh, and stuff some up the hole in the shank. Lay the ham in a trough, and cover it carefully to exclude dust and flies; baste it with the brine and turn it every day for a fortnight. Then take it out and press it on a draining board for a day.

Remove one end from, and wash out, a treacle barrel or a wooden cask. Make a heap of small birch branches and/or juniper branches and/or oak sawdust in the bottom of the treacle barrel, and bury in the heap a thick bar or lump of red-hot iron; hang the ham from the top of the barrel, and cover to conserve the smoke; punch one or two small holes in the bottom of the barrel to keep the smouldering going. Smoke the ham in this way for as near a fortnight as may be. Then hang it in the kitchen till required. It can be dried by hanging in the kitchen if necessary; but the flavour is much improved by smoking, especially if juniper is used.

Chapter Fourteen

Leather and Mohair

Goat skin provides the raw material for many of the top-quality products of the leather industry. Morocco leather is derived from the long-haired goats of cooler climates. Glacé kid and suède kid shoe leathers are derived from nine- to eighteen-month-old castrates and adult goats of warmer regions. True kidskins, from animals of one to six months old, supply a rather limited market for high-class glove making.

A kidskin when dried will weigh up to 14 oz (397 g), an ordinary goatskin from 1½ to 2 lb (680 to 907 g) and the skin of an adult male from 3 to 5 lb (1·4 to 2·25 kg). The potential output from Britain's twenty to thirty thousand goats would appear to be considerable at present-day leather prices.

Home-produced goatskin is essentially of poor quality; and, in any country, the marketing of raw goatskin is dependent on the existence of a substantial market for goats' meat. Goatskins, like other skins, are no more than by-products of the slaughterhouse.

The best skins come from goats on a low standard of nutrition; in practice, they come mainly from goats which are kept primarily for meat production in areas too poor to sustain dairy enterprise. The skin of the well-fed goat is too heavily impregnated with fat to dress satisfactorily; the texture of the skin of housed goats is weakened by abnormal activity of the sweat glands. The best goatskin in Britain is worn by the scattered flocks of feral goats, and this will make up into morocco leather of fair quality.

Nevertheless, goatkeeping is so often a commendable gesture of independence that many goatkeepers will be interested in using this goat product. Though it may prove impossible to produce a first-class article, we can produce, from our own goatskin, sound and serviceable leather goods for our own use at about one-fifth of the cost of buying comparable articles.

The quality of a skin is much influenced by the time of year at which the goat is killed. During the summer the skin of the goat is pinkish and full of small blood vessels, which are conveying into and storing in the skin the nutrients required to produce the denser winter coat. Leather made from such skin is weak and open in texture, and difficult to cure satisfactorily because of the presence of perishable nutrients which give the leather a muddy appearance. When the goat is in full winter coat, the skin is white and the skin nutrients have been transferred to the hair. At this stage, both leather and hair are at their best, and quality goatskins for use as rugs, can command quite high prices.

Skinning the goat should always be done when the body is still warm; the ligaments attaching the skin to the body of the goat are remarkably strong; if left until cold, it is extremely difficult to remove the skin cleanly. With a sharp knife, make a single clean, light cut from a central point between the two teats to the skin above the breast bone. Do not cut more deeply than is necessary to penetrate the outer skin. From the same point make a cut to the skin above the first joint of the hind legs. Loosen the skin from the belly and thighs, using the fingers and a small wooden paddle, like a flat wooden spoon, but not a knife.* Then carry on the leg cuts to a point just above the hoof. Loosen the skin around the anus and vulva, and cut around these openings. Cut a slit along the centre underside of the tail, and peel the tail skin. Now loosen and peel the skin from the back and flanks, using the paddle to loosen the skin right up to the front of the chest. Continue the first belly cut up to the throat, and work the skin carefully free of the keel of the chest, where the attachment is very close. Cut a slit up to the first joints of the front legs, and strip them in the same way as the hind legs. Cut right round the head just behind the jaw bone and ears, and peel the remaining skin from the neck; it is seldom worth skinning a goat's head.

To remove the hair from a goatskin, make a solution of 4 quarts (4·5 litres) slaked lime and 5 gallons (22·7 litres) of soft water; then stir in 9 pints (5·1 litres) of hardwood ashes. Stir the mixture and soak the skin in it for 3 hours, then hang it on the side of the tub for a

---

* The butcher uses a sharp knife, and does the job more quickly; but for skin and skinner, this is a safer method.

few minutes to drain off. Do this four times the first day, three times the next day, and once a day thereafter, until the hair parts on the thickest part; then rinse off in clean water and scrape off the hair.

A method for tanning goatskins with the hair on is as follows: Place the skin in a 25 per cent formalin solution for one week. Remove from the solution, and wash in clean water until free of formalin. At this stage, it may be possible to remove surplus tissue from the skin. Carefully stretch on a suitable-sized board, flesh side out, to dry, using nails to hold the tension (around the outer edge of the skin). When dry, rub in 'Lancrolin' oil to soften the skin, and scrape it to remove surplus tissue (this should be done with care). This step should be repeated at intervals (daily) until the skin is clean, soft and pliable. Then wash in household detergent, to remove surplus 'Lancrolin', and re-stretch to dry, hair side out.

To store skins for curing, prepare a concentrated brine solution by adding salt to boiling water until the water will take up no more; cool, and store the brine solution in airtight containers. Dip the skin to be stored in the brine solution, lay it hair side down on a wooden floor which has been sprinkled with salt; sprinkle the flesh side of the skin with salt; lay subsequent skins, similarly treated, on top of the first, in piles of up to 25 skins. When no further skins are likely to be forthcoming for a considerable time, bundle the salted skins by rolling up the pile; tie with string, wrap in thick brown paper to exclude flies and moths, and store for up to 4 months. Before curing, soak and rinse the skins thoroughly.

To cut a tanned skin to shape, lay it, hair side down, on a wooden table. Mark out with a pencil, and cut with a razor blade, or special leather knife.

Goat hair has a number of commercial uses. The coarse, long hair is, in some countries, used for weaving into tent cloth and for making the basis of carpets and rugs. Mohair, the produce of the Angora goat which is widely farmed in the U.S.A., South Africa, and, to a lesser extent, in countries of the eastern Mediterranean, is an important textile. Cashmere wool is the fluffy undercoat of goats living in the high altitudes of the Himalayas. The annual crop is hand-combed from the goat, the process taking one or two weeks to complete.

The Cashmere goat will live in Britain, but the quality of its fleece deteriorates in a mild climate; there is no economic prospect for it.

Angora goats will also live in Britain, but their fleece deteriorates in areas where the average annual rainfall exceeds 20 in (51 cm). A flock kept by the late Duke of Wellington in Hampshire produced mohair of a quality comparable to the second and third grades of imported mohair, and superior to that produced in the eastern Mediterranean.

There probably exist a few scattered areas along the east coast of Britain which could be exploited more efficiently by Angora goats than by sheep. As a farming proposition, the Angora is to all intents and purposes a sheep that will live on scrub and weeds. The fleece is heavier and worth more per pound than that of the Blackface sheep, and is sheared in the same way. The flesh is superior to all other goat meat, and at least as acceptable as prime mutton to the fat-abhorrent housewife. The milk yield has no economic importance.

As far as the hair of our dairy goats is concerned, short hair is of no interest; but there exists a very small demand for long goat hair, with top prices for white. This is used now to a very limited extent for sporran making.

Chapter Fifteen

Cropping for Goats

This chapter is intended to help the domestic goatkeeper who wishes to grow a substantial proportion of the food required by two or three milkers. If proper use is to be made of the manual labour involved, and of the restricted cropping space normally available, crop production on such a small scale calls for exceptional accuracy in estimating and applying the quantities of seed and manure needed, and in allocating space to appropriate crops.

The goat cropping ground under such circumstances is liable to be interchangeable with the kitchen garden; so the soil must be treated in a way consistent with high-quality market garden production, and not like ordinary farmland. A high humus content must be maintained, and the fertilizers used must leave no toxic residues. The need to grow field crops by garden methods, in quantities nicely adjusted to the appetite of two or three goats, is not catered for in either agricultural or horticultural textbooks. So a few guidelines are offered below.

A goat giving up to 350 gallons (1,591 litres) a year, which is as good as a goat as most of us can hope to own, needs the cropping capacity of rather more than a $\frac{1}{2}$ acre (0·2 ha) to supply all her food. A goatling needs about a $\frac{1}{3}$ acre (0·15 ha); a kid needs about a $\frac{1}{4}$ acre (0·1 ha) in its first year.

In allocating the cropping ground to different crops, and matching the size of the various plots to the goats' needs, it is convenient to use as our unit of reference the 'pole' or 'rod' of $30\frac{1}{4}$ square yards (25 m²). It is easily visualized as the standard 3 yards by 10 yards (2·7 × 9·1 m) vegetable garden bed. One hundred and sixty poles go to the acre (0·4 ha); 1 ton (1·02 tonnes) per acre is equivalent to

1 stone (14 lb) (6·4 kg) per pole (25 m²) which eases the translation of field crop recipes. In terms of poles, a good milker needs about 95 (1303 ·m²), a goatling 54 (740 m²) and a kid 37 (54 m²)—that is to grow all its food, concentrates, hay, the lot.

To calculate the goats' needs in terms of actual crops, winter and summer feeding must be considered separately. In winter, a 180 lb (81·6 kg) goat in milk will need about 5 lb (2·25 kg) of hay a day, to maintain health and butter fats, her standard 2 lb (0·9 kg) of concentrates for maintenance, and as much kale and roots as her milk yield justifies. In terms of starch equivalent (s.e.), that is 1½ lb (680 g) s.e. from hay, 1¼ lb (567 g) s.e. from concentrates and 2¼ lb (1 kg) s.e. from kale or roots, for a goat giving 1 gallon (4·5 litres) of milk a day. Winter is shorter in the south than in the north, but assuming a 6-month northern winter, the goat is going to need a total of 270 lb (122 kg) s.e. from hay, 225 lb (102 kg) from concentrates and 405 lb (184 kg) from kale etc.

In summer, we may allow the same 6 month's concentrate requirement of 225 lb (102 kg) s.e.; the rest of the ration, say 675 lb (306 kg) s.e. will be wanted in the form of fresh green fodder.

The productivity of land, in terms of starch equivalent, varies somewhat with inherent fertility, and the suitability of crop to soil and climate; but, on garden-size plots worked by garden methods, we can assume a high level of fertility; provided we choose crops to suit the district, we can make some useful generalizations. Hay and grain crops yield about 14 lb (6·4 kg) starch equivalent per pole (25 m²); green crops, such as grass, cabbage, kale and lucerne, yield about 20 lb (9 kg) starch equivalent per pole, and enough protein to balance the starch equivalent for milk production. The popular root crops yield about the same amount of starch equivalent as the green crops, but with very little protein. Table 12 summarizes the position.

Of course, neither milking goats, kids nor goatlings maintain this conveniently level appetite throughout the year. But, when planning the cropping, it is impossible to forecast food needs with day-to-day accuracy months ahead. The method suggested here maintains a flexible link between supply and demand, and includes a necessary margin for error and for partial crop failures. Many domestic goatkeepers may prefer to buy in their concentrates, or their hay, or both; using this method, it is easy to adjust the cropping

Table 12: Area requirements of goat cropping ground

	per day		per 6 months		in cropping ground	
	lb s.e.	g	lb s.e.	kg	poles	m²
Needed for a milker:						
(winter)						
from hay	1½	680	270	122·5	20	506
from corn	1¼	567	225	102·0	16	405
from kale etc.	2¼	1,020	405	204·1	20	506
(summer)						
from corn	1¼	567	225	102·0	16	405
from green crops	3¾		675	306·2	34	860
Needed for a kid:						
(1st summer)						
green crops	1½	680	270	122·5	13	329
(winter)						
from hay	1	454	180	81·6	13	329
from corn	½	227	90	40.8	6	152
from kale etc.	½	227	90	40·8	5	126
Needed for a goatling:						
(winter)						
from hay	1½	680	270	122·5	20	506
from corn	½	227	90	40·8	6	152
from kale etc.	½	227	90	40·8	5	126
(summer)						
from green crops	2½	1,134	450	204·1	23	582

plan accordingly. The productivity of the major goat crops is listed in Table 13.

There is no great difficulty in growing the goats' corn and concentrates in small, garden-scale plots, but special considerations are involved. In the British climate the choice of high-protein concentrate crops is limited, practically speaking, to beans, peas, and linseed. Goats are not always very keen on beans, which have a slightly constipating effect. Peas are tricky in a wet season in a wet district. Linseed, too, prefers a sunny climate, and suffers much damage from wet harvest weather; even in the best of seasons, it is not a heavy cropper, yielding only 8 lb (3·6 kg) of starch equivalent per pole (25 m²) of a valuable, but not very convenient, food. In

Table 13: Productivity of major goat crops

Crop	Weight of crop per pole		s.e. per pole		Notes
	lb	kg	lb	kg	
Hay	30–40	13·6–18·1	10–14	4·5–6·3	
Oats (grain only)	18	8·2	11	5·0	
(straw only)	28	12·7	5½	2·5	
Barley (grain only)	21	9·5	14	6·3	
(straw only)	21	9·5	5½	2·5	Seldom eaten
Beans (seed only)	28	12·7	14	6·3	Protein-rich
(straw only)	31	14·1	5½	2·5	
Peas (seed only)	28	12·7	14	6·3	Protein-rich
(straw only)	30	13·6	5	2·3	Edible
Linseed	7	3·2	8	3·6	Protein-rich
Lucerne (hay)	40	18·1	11	5·0	Protein-rich
(fresh-cut)	210	95·2	21	9·5	Protein-rich
Cabbage (drumhead)	240	108·9	14	6·3	Protein-rich
(open-headed)	210	95·2	16	7·3	Protein-rich
Kale (marrow-stem or thousand-headed)	210	95·2	19	8·6	Protein-rich
Maize (cut when cobs are milky)	180	81·6	17	7·7	
Comfrey	300	136·1	15	6·8	Protein-rich and very early
Buckwheat	100	45·3	8	3·6	Very tasty, for poor ground
Chicory	200	90·7	10	4·5	Very early
'Herbal lea' (cut)	180	81.6	24	10·9	Cut when under 1 ft (30 cm) high
(strip grazed)	150	60·0	21	9·5	
Italian ryegrass (cut)	200	90·7	22	10·0	
Nettle hay	35	15·9	14	6·3	High-protein
Fodder radish	200	90·7	16	7·3	
Rape	180	81·6	12	5·4	
Fodder beet	150	60·0	19	8·6	
Potatoes	110	49·9	20	9·1	Feed baked or boiled
Mangolds	300	136·1	18	8·2	
Swedes	240	108·9	17	7·7	
Carrots	130	59·0	12	5·4	

many districts it may prove advisable to grow the concentrate ration in the form of oats or barley, and to rely on the fresh green foods to provide the protein needed to balance the diet. On the other hand, if a balanced concentrate mixture is bought in, it may be preferable not to stake too much on the kale crop, which is the key to winter protein; kale too often falls victim to accidents of weather and the persistence of wood pigeons. Fodder beet and potatoes are much more reliable crops for winter feeding. In summer, a succession of mashlum (p. 302) maize and fodder radish can replace lucerne, with a gain in variety of diet and a saving of cropping space—provided the concentrate ration is a bought-in, balanced mixture.

If grain crops are to be grown to provide the concentrate ration, their straw will relieve the demand for hay—if their straw is edible. No allowance is made for this uncertain factor in the table above. By harvesting the cereal crop before it is quite ripe, and feeding it 'in the sheaf', the hay and concentrate ration may be combined—which has advantages for kids and goatlings, and disadvantages for heavy milkers.

Are the goats to be put out to graze during the summer? It will save a lot of labour if they can thus be made to do their own food gathering. But it is very difficult for a few goats to make full use of a small area of ground if they are put out to graze on it. Five goats, put out to graze on 1 acre (0·4 ha) of pasture which is the minimum needed to provide their summer keep, concentrates apart, would be faced in July with a patchy paddock of seeded and exhausted grasses, carrying a burden of parasitic worm larvae which would explode into big trouble with each spell of warm, wet weather. This is the most uneconomic way to use pasture. Divide the acre into four and rotate the goats around these four little paddocks, allowing about a week in each, and the results would be better; but much grass would be fouled with droppings and wasted. Strip grazing, with the goats on running tethers (p. 82) or controlled by electric fencing, makes good use of pasture; but the goats may become so fretful with the close restraint, especially when discomfited by hot sun, flies, wind or rain, that they fail to eat their fill, and fail to give their best. With five goats to feed, the balance of advantage may lie with rotational or strip grazing. With two or three goats and similarly restricted acreage, it is often best to cut and carry all they require—in which case, grass, though indispensable at some seasons, is not always and everywhere the ideal crop.

For such crops as are normally grown in the garden, gardening books will provide a wealth of information on the manuring and methods of cultivation; for some of the other crops listed in Table 13 the following notes may be useful.

OATS

These yield better than barley in wet, cold, districts, and provide a more balanced feed than barley anywhere. In wet, late districts a soft-strawed oat, such as Bell or Castleton, is a good proposition, especially if cut on the green side and fed in the sheaf (i.e. unthreshed). In drier districts a grain oat, with a relatively hard, short, straw is generally preferred; Sun II is a good dual-purpose type.

MANURING:

1 barrowload of compost (not dung) per pole (25 m²), 5 lb (2·25 kg) hoof-and-horn meal, and 5 lb bone meal, worked into the seed bed.

Sow 12–14 oz (340–397 g) per pole in March to April, in seed bed with moderate tilth.

Cut, if required green, when the grain exudes milk when pressed; if cutting for grain, cut when the last tinge of green is disappearing from the heads, and leave in stocks or on tripods to dry for a fortnight.

BARLEY

Manuring: 1 barrowload of goat dung, 7 lb (3·2 kg) hoof-and-horn meal, and 4 lb (1·8 kg) bone meal, worked into the seed bed.

Sow 12–20 oz (340–567 g) per pole (25 m²), according to the coldness and wetness of the district and the lateness of the sowing, into a fine, deep seed bed from early April onwards.

Cut for hay or green fodder in the flowering stage; cut for grain when dead ripe.

BEANS

Use field beans for a seed crop (though runner beans are a productive green crop).

Manuring: 2 barrowloads of goat dung, and 5 lb (2·25 kg) of bone meal per pole (25 m²) into the drill in late autumn; or spread, for a broadcast drop, in spring.

Sow: 1½ lb (680 g) of seed per pole at 3-in (8 cm) intervals, in the dunged drills 20-24 in (50–60 cm) apart, in February.

Cut when the middle pods are just ripe, and dry on tripods. If to be grown mixed with oats, as mashlum, broadcast 14 oz (397 g) beans per pole into the manured seed bed in February, and 12 oz (340 g) of a soft-strawed oat 3 weeks later. Cut when the beans are ready. Dry on tripods, set over paper sacks to save the seed.

PEAS

Manuring: 2 barrowloads of compost (not dung) with 7 lb (3·2 kg) bone meal per pole (25 m²).

Sow 3 oz (85 g) of peas with 12 oz (340 g) of oats, or 4 oz (113 g) of peas with 10 oz (283 g) of oats, in a deep, moderate tilth in mid-March.

Cut late August for grain, at the oat-milk stage for hay, or when the first pods are full for use as green fodder. Use tripods for drying.

LINSEED

Manuring: a light dressing of compost, or none at all if the previous crop was well manured.

Sow 6–7 oz (170–198 g) per pole (25 m²) in a deep, fine tilth, in April.

Cut when the first bolls ripen, using a sharp blade and leaving a 6-in (15 cm) stubble; the stem is very tough close to the ground. Linseed must be fed freshly ground, preferable mixed with bruised oats. The crop can be matured in the stock or on tripods; there is always some loss of seed and, as young green linseed is a valuable and tasty feed, it is well to follow linseed with a green forage crop, such as peas and oats.

LUCERNE

Manuring: lucerne is a glutton for dung and compost throughout its long life. Use 2 barrowloads of compost per pole (25 m²) in the autumn before sowing, 1 barrowload of dung worked into the seed bed in spring with 7 lb (3·2 kg) of bone meal, then 1 barrowload of compost per pole (25 m²) per year.

Sow: 2 oz (57 g) per pole, ½ in (13 mm) deep, in drills 10 in (25 cm apart, allowing 2 oz of seed per pole, in April. Soak the seed in bacterial culture, obtainable from the seedsman, if the ground has not recently carried lucerne.

Cut before the flowers fade.

COMFREY

Manuring: comfrey responds well to dressings of dung and compost and hoof-and-horn meal, once it is well established; but in its first year in a fertile soil, no manuring is required; on a hungry patch, 1 barrowload of compost per pole (25 m²) will be sufficient.

Planting: plant the sets 3 ft (0·9 m) apart, both ways, in ground cleared of perennial weeds, in April. If the sets are not fresh and vigorous, keep them moist in a box of light compost in a closed frame until new growth is started, before planting them out.

Cut to within 3 in (8 cm) of the crown of the plant, as soon as flowering stems appear, or at any time when the plant exceeds 18 in (45 cm) in height. Do not let the plants flower, even if the leaves are not needed immediately for feeding; use unwanted cuts as a mulch or as a compost accelerator.

CHICORY

Manuring: chicory is not a greedy feeder, but responds well to dressings of compost applied between the drills in early spring. Work 1 barrowload per pole (25 m²) of compost into the seed bed, with 5 lb (2·25 kg) of hoof-and-horn meal, and 5 lb of bone meal.

Sow 2 oz (57 g) per pole, ½ in (13 mm) deep, in drills 1 ft (30 cm) apart, in early May. The plants may be thinned to 9 in (23 cm) apart, but this is not necessary unless the plot is to be robbed of chicory roots for forcing.

Cut before flowering for green fodder; in the flowering stage for hay.

'HERBAL LEA' MIXTURES

For a grazing mixture, buy a small quantity of a standard herbal lea mixture from a recommended supplier (p. 362).

The proportions suggested are as follows (weights are *per acre* [0·4 ha]):

8 lb (3·6 kg) perennial ryegrass
10 lb (4·5 kg) cocksfoot (or 8 lb [3·6 kg] timothy on heavy soil)
1 lb (454 g) rough-stalked meadow grass
3 lb (1·4 kg) late-flowering Montgomery red clover
2 lb (0·9 kg) New Zealand mother white clover
1 lb (454 g) Kent wild white clover
1 lb (454 g) alsike

2 lb (0·9 kg) chicory
4 lb (1·8 kg) burnet (omit on acid soils)
2 lb (0·9 kg) lucerne (omit on acid soils)
1 lb (454 g) plantain
16 lb (7·3 kg) Italian ryegrass

Five oz (142 g) per pole (25 m²) of this sort of mixture will suffice; or 3½ oz (99 g) without the Italian ryegrass which is sown as a temporary cover crop. Seven oz (198 g) per pole of oats might be substituted for the Italian ryegrass.

For a cutting mixture, something simpler and cheaper will give equally satisfactory results (weights are per pole [25 m²]):

1½ oz (42 g) cocksfoot (or 1 oz [28 g] timothy on heavy land) hay
strain
½ oz (14 g) late-flowering Montgomery red clover
½ oz (14 g) lucerne
½ oz (14 g) chicory

A little wild seed of plantain and dandelion may be added to this with advantage, provided the neighbouring vegetable gardener has no objection to the appearance of a few dandelion seedlings.

Manuring: 2 barrowloads compost per pole, with 7 lb (3·2 kg) hoof-and-horn meal, and 7 lb bone meal, into the seed bed; 1 barrowload of dung to be spread each winter after the lea is established.

Sow in early April or July on to a fine, firm seed bed on a calm, dry day; use the seeding rates given above; roll or tread the seed in, and water it in with a fine rose, lawn sprinkler or hose spray, but do not cover the seed with soil. In April, it may be necessary to offer an alternative supply of bird seed in a convenient container nearby; in July there are ample suppies of bird seed in every field and garden. *Cut* the crop when it is 8 in to 1 ft (20 to 30 cm) high, for green fodder; cut it in the flowering stage for hay.

KALE

Cultivation and manuring as for cabbages. Thousand-headed, 'Canson' and marrow-stem kale are sown in April and crop from September to February; asparagus kale is sown in April and crops from March to May; 'Hungry Gap' kale is sown in July and crops from March to May however hard the winter may be.

BUCKWHEAT
Manuring: none.
Sow 12 oz (340 g) per pole (25 m²), for use as green fodder, in the second week of May. It is very palatable. It may be sown, 6 oz (170 g) per pole, as a nurse crop for 'herbal lea'.

NETTLES
These may be grown in bottomless buckets submerged to their rims in the soil, or in a bed edged with 1 ft (30 cm) wide corrugated-iron strips (e.g. Nissen-hut link strips), set their width into the soil.
Manuring: dung and coal-fire ashes.
Cut in the flowering stage.

FODDER RADISH AND FODDER BEET
These require much the same treatment as the related vegetable. Fodder radish is a green fodder catch crop, recently introduced as a big improvement on rape, being more productive, more nutritious and free of the photo-sensitizing effects of rape. It is usefully sown any time from early May to the end of August.

The harvesting of small-scale goat crops can be made easier if some form of drying rack is used to dry such crops as comfrey and chicory and lucerne and grass cut at the green-fodder (under 1 ft [30 cm] high) stage. A poultry field ark, with a slatted or wire-mesh floor, raised 1 ft off the ground to create a good under-draught, performs the function well. A purpose-made ark, using black corrugated-iron sheets to absorb the maximum of sun heat, and providing generous ventilation at the ridge, would probably do the job better. Palatable silage, with less stench than the normal, can be made in plastic bags; fill the bags with the wilted greenstuff, extract the air with the suction pipe of a milking machine, and seal. This is an entirely practical and economic proceeding, but only if you are scrupulously careful in handling the plastic bags, so that they can be re-used. A vacuum cleaner would take a little longer than a milking machine, but if the material in the bags were not too coarse, the vacuum cleaner would be a good tool for silage making.

CALCIFIED SEAWEED
This is a coral-like substance, which is useful both as a fertilizer and as a food additive. It is carried by the Gulf Stream and found in huge

deposits off the coast of Brittany. From there, at a depth of a hundred feet, it is mined and finally arrives in our warehouses and garden shops in the form of a fine white powder, variously known as Seagold, Mermin, etc.

It contains an impressive quantity of calcium and magnesium as well as a wide range of trace elements and, because of its slow action, is safe to apply to the ground, having none of the disadvantages of lime. Farmers using it in all parts of Britain report an improvement in soil structure and fertility as a result as well as an all-round increase in the health and productivity of beef and dairy animals. When one farmer included it in his dairy ration, it saved him from having his milk rejected.

This product is used as a general fertilizer on crops and grass, on farms and gardens. It is included by merchants in some dairy feeds, and a sprinkle of it—about 1 oz (28 g) daily—for a dairy goat on concentrate feed may help butter fats. It can also be useful for the vegetable garden and smallholding. Although it is expensive to buy, it is so finely ground that it is economical to use.

Chapter Sixteen

Goat-farming Systems

To illustrate the operation of principles set forth in previous chapters, here follows a series of descriptions of goat-farming enterprises. These are largely written by the owners themselves, and consequently reflect their own experiences and views, which may therefore differ a little from those expressed in the foregoing part of this book.

THE HOUSEHOLD GOAT

This is an example of a small, domestic enterprise run as a sideline—but quite a professional one. I am a professional forester living near Aviemore, Inverness-shire; and I keep two milkers and a kid, feeding them partly from natural vegetation and partly from my vegetable garden. This has an area of 180 sq yd (150 m²) but is mainly for kitchen use, though with a large section of kale and turnips for the goats. Pea and bean haulm and pods are fed to the goats as are all other vegetable trimmings.

As I live at the edge of woodland, I have two enclosures of ⅓ and 1 acre (0·15 and 0·4 ha) (fields would be too flattering a title), both with goat sheds. These I used in rotation for a few months at a time. Their main vegetation is mature Scots pine with a ground layer of heather, blaeberry, grasses, etc. The goats seem particularly fond of the old pine bark, and lichens growing on it.

I have the main goat shed in the garden beside the house; and here the goats spend most of the winter, apart from daily exercise. My house is 1,100 ft (335 m) above sea level, at the foot of the

Cairngorms. Each shed is approximately 7 ft × 9 ft (2·1 × 2·75 m) based on the British Goat Society leaflet *Plans for a Small Goat House.* Hay and other feeding stuffs are kept in a separate shed.

The goats are British Toggenburgs; my original aim was to have only the two milkers, though at present I have one kid in addition. I breed them alternately, biennially; and I always go to a pedigree B.T. or pure 'Togg' for stud purposes. As I am trying to upgrade and improve my stock, I expect I shall always keep a kid for rearing. Surplus female kids are sold as soon as possible. Male kids are killed at birth, as I value the milk more than the meat which would be used to produce it.

A block of standard Rumevite is always on the wall of the goat shed. Oats and bran are fed at each of two daily feeds, as well as kitchen scraps, roots and some greens. Hay is fed all the year round and water (warm in winter) changed twice a day. During the summer months, the goats are tethered during the day, involving about four shifts, while the fields are being reseeded. Branches are fed as and when available. Scraps from a local greengrocer and a youth hostel are utilized when obtainable.

I place great emphasis on strict routine:

6·45 a.m.   Feed, water and milk
12·00 noon   Check hay, give roots or any titbits depending on the season
6·00 p.m.   Feed and water
7·00 p.m.   Milk

Yield averages 385 gallons (1,750 litres) per year for two goats. I usually have regular customers for 2 pints (1·1 litres) daily. My household consists of myself, my wife, and two children aged three and five, requiring 4 pints (2·3 litres) daily. Any surplus is sold to campers from a nearby site (surplus outwith the busy camping season is deep-frozen for future use). Milk is sold at 10p per pint in customers' own containers.

I make my own hay from 'wild herbage'. In practice I utilize picnic areas, open areas near the forestry campsite, roadsides, old estate gardens, etc. The goats are particularly fond of nettle and raspberry-leaf hay.

I obtained my original goat free and, being something of a handyman, made my own sheds and trailer, In practice, the trailer is

used for firewood, shopping, etc., so its cost is not wholly account-
able to goats.

Dairy equipment consists mainly of normal kitchenware:
stainless-steel saucepans (for milking pails), sieve and paper nappy
liners for strainers, plastic containers for storage in the fridge. Here
again the cost is hardly attributable to goats when the equipment is
there anyway.

Replacement goats come from my own stock.

Expenses:

14 cwt (711 kg) bruised oats and bran	£47·76
2 blocks Rumevite	4·46
Seeds	1·00
Service fee, registrations, transfers, etc.	9·40
Vet's fees and drugs, including disbudding	6·00
	£68·62

(The total doesn't include transport, as this is part of the family
hobby)

Income:

Sale of one female kid (on average)	£10·00
Milk sales (115 gallons [523 litres] @ 10p per pint)	92·00
	£102·00

(80 gallons [363 litres] of milk are used for kid rearing, and 190
gallons [863 litres]—value £152 at the above rate—go for home
use)

BRYCE REYNARD

GOAT FARM IN WEST WALES

This is an example of an 80-acre (32·5 ha) farm stocked mainly with
goats, but with some cattle as a sideline. The farm overlooks Car-
digan Bay, only a few miles from the sea but about 600 ft (150 m)
above sea level. The climate is mild and inclined to be showery;
heavy rain and snow are rare. The land was desribed as 'not of the
best agricultural type' when purchased seventeen years ago; since
then, draining and reseeding have been done with Ministry grants,
and rushes are no longer a feature. The clay subsoil is not far from
the surface, and water takes some time to drain off many fields,
making poaching a problem; but generally grass continues to grow

for much of the winter, and good crops of hay and silage are harvested. Ditches and open drains surround the fields, which are mostly of 1½ to 3 acres (0·6 to 1·2 ha) and surrounded by hedges of beech and blackthorn.

In addition to the goats, the farm has been stocked with a beef herd of cattle-suckler cows rearing their own calves. The farm qualifies for the Hill Cow subsidy, which has greatly helped to improve conditions. It is managed with an adjacent dairy farm by the one owner, although both farms are separately staffed and stocked, with a shared relief. A girl is employed for rearing young stock and for the goats.

Originally, the main business of the farm was cattle, and it was the goats that were the subsidiary enterprise, so that income from the goats is not easy to establish; but milk and produce have been sold for a number of years, during which it became increasingly apparent that the goats were a worthwhile proposition in their own right. Knowledge had already been acquired of the type of housing, handling numbers, grass seeds to grow, milking arrangements and so on; and the growing demand for goats' milk, cartoned or as yoghourt, encouraged the decision to make the goats the main farming interest. The return of the owner's son to take on the dairy farm and 200 acres (81 ha) of the land, leaving the stock farm with 80 or so acres (32·5 ha) to carry goats and the remains of the Hereford herd—perhaps ten or twelve animals—has coincided with a need for reappraisal, since the large increase in costs, especially for labour, renders marginal farming unrewarding, unless family owned and managed.

From previous costings, a milking herd of 300 goats is eventually envisaged, producing cartoned milk for a market or supplying yoghourt to a distributing firm. At present, up to 60 animals have been milked, and the numbers will grow within the herd. The costing details of the proposed system have been worked out by the Milk Marketing Board's L.C.P. service.

To farm this type of land with 300 goats requires a fairly intensive system. Constant daily free range, as at present enjoyed, will only be possible in dry weather after grass cutting, and must be regarded simply as exercise. The empty cattle yards will easily accommodate the milkers, allowing scope for kid yards and male accommodation. The (herbal) grass will be cut and fed in troughs and racks after slight conversion. Rye and other green crops which do well in this part of the country are already grown for feeding in racks. Home-produced hay can, if necessary, be supplemented locally. Straw must be bought in, also beet pulp and dairy rations; oats are purchased in bulk and rolled as required.

Apart from the (cattle) yards, and the smaller feeding space for every animal, there is the milking parlour, which has pipeline equipment, made by Alfa-Laval, for four animals milked simultaneously, with the machine adjusted for goats. Otherwise, equipment and facilities are similar for both herds, as are the standards of cleanliness and tidiness around the parlour, since the dairy must be open to inspection at any time. Individual rations are fed in the milking parlour. General feed and hay, unrationed, are in the troughs.

Kids

As far as possible, the female goats are batch served, so that kiddings occur in groups. The service list is made up in the summer. It is arranged for the first group to kid early in the new year, by utilizing aids available for planning seasons, as with ewes.

Kids stay with their dams for four days; then the dam goes out with the herd, and the kid joins a group of young kids. Each kid is earmarked, disbudded where necessary—under general anaesthetic—and well examined. In the afternoon, the feeder bucket, with teats as for lambs and holding substitute lamb milk, gets each kid sucking. A small group is kept together until kids approach the lambar eagerly and take a teat; then the group will be made up to twelve kids. Hay and water are provided, and a protected feeder hopper supplies the ration—rolled oats and flaked maize, which are frequently replenished. When they are eating well, a coarse protein ration is added. Amounts increase, and the progress of each animal is noted. At two months the kids are still getting substitute milk from the lambar feeder three times a day. When they are judged ready, the lamb-milk substitute is gradually replaced by calf-milk substitute, which continues until the kids are over four months. By then, the quantity of dry food consumed is adequate, taken with hay and water, so the milk feed can be gradually dropped.

It is reckoned that the kid at six months weighs 50 to 60 lb (22·5 to 27·2 kg) and will have cost £25 to feed. Generally the kids do not go from the open yards. Female kids to be retained are kept together; male kids considered worthy and required for stud purposes are reared in a similar fashion, but continue with substitute milk for as long as they will take it. Two or three are kept together. Unwanted kids are killed at birth. Any to be reared for meat are castrated, run together and weighed frequently.

Before Christmas female kids are examined; and those that qualify in size and maturity will be served by a male kid, these served kids going out for exercise with the adults, but receiving extra rations and being penned together. Before kidding, they will have become used to the parlour by going in and finding rations. The goat must jump on to a raised bench and put her head through a gap to find the bucket. It is found that the newly kidded yearlings enter the parlour and accept the milking machine without trouble, after this practice.

Kidding at fifteen or seventeen months, the animal will have cost £50 to £80. Thus, female kids produced in May and June are usually sold, since they will not be mature enough to breed in the first winter, and the chance of their holding to service in March is uncertain.

The system of serving kids is vital to a commercial farm: early kids do better; but, after March, progress seems to take longer. Every way must be sought for cutting costs, not by skimping on feed, but by eliminating wasted food and labour. The animal must be productive as soon as, and as long as, satisfactorily possible. Goats are unlikely to milk economically after ten years of age, so approximately nine years' milking life must pay for rearing and yearly keep, as well as the overheads—there is no carcase value at the end.

Milking Goats

Breed is immaterial; milk potential for long lactation, hardiness and good conformation with well-attached udder are the important factors. The highest yielder is not necessarily the best commercial proposition; an average medium milker gives over 1 gallon (4·5 litres) a day in the first half year, and just under for the rest. This taxes her not at all; and she will give 300 gallons (1,364 litres) for each of her milking years, and a bit more in her prime, without loss of condition. With modern aids for planned seasons, groups of goats could be kidded 'out of season' as a regular system to ensure a constant milk supply from the herd. This may be a future practice;

but at present, with care, it is possible to keep a regular supply by running through half the milkers each year.

Males

On this farm the policy is to buy in stud males as kids, with a record of milk and good conformation behind them. The few which are kept from the herd kiddings and the bought ones are reared together, going into adjacent single pens in late August. The kids get well fed throughout their first winter; but dry feed is preferred to continuing the milk substitute, so milk powder is introduced in small amounts in their feed if it should appear advisable. Male kids are used to serve the small females and kids, a few services each.

At the beginning of April, the five or six adult males and the bucklings are put out on a safe paddock with shelter, water, a hay rack and a trough under cover. Since the field is never full of lush grass—for obvious reasons—by the time it grows, hay and feed are refused. They live out all summer, and it is one of the happiest sights on the farm: a gathering of noble-looking gentlemen lying in the sun, with the young lads gathered about them. As soon as precedence is established, they give no trouble at all.

The bucklings come into their winter quarters during August, not, as one might imagine, because they could get hurt by aggressive older males, but because they tend to gang up on their elders as the season approaches: we have had a punch-drunk adult male who most certainly got his condition from two sparring young ones getting his head in between theirs. The older males stay out until September, if the weather is dry.

Housing and Yards

Communal shelter is advantageous, saving bedding, labour and space. The goats lie on raised platforms—a type of kennel/cubicle layout. Straw is used, but very economically; and the 'dung passage' is easily kept clean with a squeegee—a blade of stiff rubber instead of a broom head. Water troughs with ball valves are installed in each yard; they cannot be fouled, and the goats drink plenty from them. It is essential to have adequate room for all together at the hay manger, and that the systems for supplying hay ensure no waste. Minerals in powder form are on offer.

In this arrangement, most of the goat accommodation is in open yards, and health is good. They all have adequate shelter from wind and rain.

Milking Parlour, Dairy for Yoghourt and Cheese Making

It is important to get the goats milked quickly, and the milk cartoned and frozen. This is only possible with a milking machine and bulk refrigerated tank. In fact, we do not believe goats' milk should be sold without this guaranteed way of cooling to below 45°F (7°C) within a very short time. The milk goes directly to the bulk tank from the goat, and chilling commences immediately. As soon as the goats are released, the attendant is free to carton the milk and place it in the freezer. It is believed that milk costs 50p a gallon (4·5 litres) to produce with the herd at its present size.

Owing to the situation of this farm, markets are some distance away. For this reason, we have a cheese room arranged, and yoghourt is made in a special dairy. Converting milk into yoghourt is worth while; into cheese, less so. The best market is cartoned milk: since the costs of cartoning are minimal, the price can be realistic if it is sold on the yard. A certain amount is collected, and some is delivered with yoghourt and cheese. In summer, local agricultural shows are well worth support, and a great deal of produce is sold, which leads to further orders. By far the best outlet for any goat enterprise must be to sell at the farm to a distributor.

Labour

Labour needs are usually said to be excessive in handling goats: but day-to-day management need not be so. Where most goat owners go wrong is in expecting goats to behave like other stock. Provided the eccentricity of the goat is understood, and the owner thinks one step ahead, problems are few. At a certain time in the afternoon, the herd presents itself at the field gate nearest the farm, waiting to come in; and if, for some untoward reason, their attendant does not arrive on time, they can hardly be blamed for finding a way home on their own.

The routine feeding and milking of possibly 300 goats may produce labour problems. To ensure an acceptable time lapse between milkings, it will be necessary to have one attendant milking from

6 a.m. and doing the morning routine; then another relieving her at noon to do the afternoon routine and evening milking. A third attendant would have to relieve the other two, as well as doing the dairy work. All this is in the interest of the 40-hour week; but gives a labour outlay of well over £100 a week, plus the little overtime pay that is necessary.

Conclusions

These are difficult to give. The goat has not often been considered a farm animal, especially in Britain; and it is possible that the farm referred to here is the first which will be officially costed. As in all livestock enterprises, the thin margin between profit and loss can be influenced one way or another by as little as the use of half a bale of hay where a quarter is needed, so that a bale and three-quarters a week is wasted.

PAMELA GRISEDALE

ZERO-GRAZING

While the notion of 'zero-grazing' goats may not appeal to advocates of a natural way of life, this example shows how it may be achieved successfully and humanely. This is no 'factory farm'.

Housing and Yards

To zero-graze successfully, you need a box stall or pen 5 ft × 7 ft (1·5 × 2·1 m) for each goat, and a big exercise yard. The pen should have a hay rack 2 ft (0·6 m) square, with a lid, so that the goat does not take the food out of the top, and thus waste half of it. There should also be a salt-lick holder with pure salt block, and two rings for buckets—one that falls flat, for the water bucket, and the other firmly fixed, so that the feed bucket can be put in and also taken out for washing. If the feeding bucket is rigid, you can put whole roots in it, thus keeping the goats occupied longer. An idle animal is often an unhappy one, and cannot do her best. Stale bread and kale stalks are eaten well in a fixed bucket; in fact you will find that goats take things back to their fixed bucket after dropping them.

If the sides of the pen are 5 ft (1·5 m) high, few goats will jump out; but, as they all like to see what is going on, cut a 'window' in the door so that they can put their heads out easily. This stops them from standing on the door and, in time, wrecking the hinges. Board the pens up to 2 ft (0·6 m) from the floor to cut out draughts, leaving a 2-in (5 cm) gap between the higher boards so that the goat can see her neighbour; but make sure that the feeding bucket, which may be in the corner, is shut in. Nothing puts a goat off its feed quicker than to have its ears nipped by a neighbour while eating.

For the floor, nothing is better than concrete sloping towards a drain. On this floor you can have wooden slats, which must be made of durable wood, such as elm; they should be in 2 in × 1 in (5 × 2·5 cm), set ½ in (1·5 cm) apart and nailed on to 4 in × 2 in (10 × 5 cm) crosspieces. These slats are heavy, and get heavier when wet, so it is as well to have them in small sizes—say, four to a pen. Slats are much better for large udders to rest on than cold concrete; also, you don't need so much straw.

It is as well to have at least two yards for the milkers so that you can separate the ones in kid, or for illness. Again, the floor should be concrete, for easy cleaning, slanting to a drain. Concrete should be roughed; otherwise it will be slippery in frosty weather. If you are building from the start, face the goat house door to the south, so that it gets all the sun and the goats are out of the north wind; if possible, build the barn on the east side to keep the goats out of the east wind as well. It is these small details that put extra ounces into the milking bucket.

A branch rack is essential in the yard; it should be made very sturdily to carry big branches that the goats can bark during the winter—and you can burn these as logs the following year. For eight goats, a rack 8 ft × 3 ft (2·5 × 0·9 m) wide, 2 ft (0·6 m) off the ground and 5 ft (1·5 m) high is needed. Have the sides made vertical; then the goats must put their heads in to eat and cannot just walk along pushing others out.

The yards can be fenced with chain link or 'Weldmesh' or, if you live near a sawmill, with straight elm boards. A shelter from the rain and hot sun is essential: without this, you cannot let them out in all weathers; and they should get out once a day to take in vitamin D through the action of the sun on their skin.

The goat house can be any well-built structure, but it is best to have a fairly high roof to trap warm air; a slate roof with thick insulating boards for a ceiling does well. Provided the pens face a 5 ft (1·5 m) wide passage, there will be no need for more than the bottom half of a stable door as entrance to the house. This way it will be neither too hot nor too cold; a goat is a tough animal and will stay tough unless coddled. Finally, have a large gutter round barn and goat house to catch all the rainwater; this can be stored in tanks with a pipe through the wall of the barn or food room. A tap with a piece of hose attached is used for filling the boiler to supply hot water. Like humans, goats prefer their drinks hot: not only do they make more milk, but there is no waste of precious food for heating the water in their bodies.

It is better for the male if he can see his females; so build his pen on the end of the goat house. He needs a large house 10 ft (3 m) square, and a run as long as the other yards. If the kids are housed between his yard and the milkers, he will always have company, and be quieter and easier to handle. Make a feed passage next to his pen so you need never go inside, getting clothes tainted with his smell and contaminating the milk. Have two bucket rings—both fixed, as he will play with them otherwise—with a large salt lick between. The hay rack should be inside his pen, with the back made as a door for filling from the feeding passage. In order to tidy up the yard without his assistance, have a door to his pen, which you can shut from outside; this is arranged by means of a steel bar attached to a sliding door; In the house, instead of littering the whole floor, it is more convenient to have two strong beds of straw-covered slats—one either side, so that he can get out of the wind. These need sides to hold the straw. As long as he is allowed to go in and out at will, the beds will remain dry—which, of course, keeps him much cleaner. A slatted bed in the yard encourages him to lie on that instead of on the wet, cold concrete. He may be found lying out even on frosty nights.

Kids also want a bench outside; and they, too, may lie out in frost. They need a lot of things in their yard for exercising: old water tanks, big rabbit-hutch-sized boxes, and an old tree trunk—if you can get it there—does well. The yard must be big so that they can race about and develop without putting on surplus fat.

Yarding goats is another form of zero-grazing, the only difference being the goats all live together, probably going in and out at will. Pens are arranged for shutting them up during kidding or illness. They have a large communal haystack but are tied at milking to eat concentrates. High yields need extra rations, so individual feeding is essential.

Feeding

If really high yields are your aim, you must study the goats and their comfort carefully. Let them out for three hours in the morning, while you clean out and refill buckets and racks. Given shelter and branches they will be quite happy for that length of time. To obtain the maximum amount of milk, feed little and often: it is useless to stuff racks twice a day and give them a bucket of cold water.

This is the scheme I use for a day's feed in a zero-grazed herd:

6·45 a.m.	Concentrates during milking; after milking, racks filled with kale in winter, herbal lea in summer.
9·00 a.m.	Milkers let out to large buckets of hot water and branches in racks.
12·00 noon	Milkers let in to hay in racks and roots in buckets (or mangers) in winter, herbal lea in summer.
4·00 p.m.	Racks filled with kale in winter, herbal lea in summer.
6·45 p.m.	Concentrates during milking; followed by hot water buckets, which are left with them for the night; kids are also offered hot water at this time; hot soaked beet pulp is given all year round.
10·00 p.m.	Racks filled with hay in winter, herbal lea in summer.

To make a good milker you must start the day she is born. She wants milk up to six months, rising to 4 pints (2·3 litres) at about a month old, tapering off towards the sixth month. Kids eat sweet, short grass before anything else solid and should have it in their racks at about ten days old. If they are born too early for grass, an old kale stalk with lots of little leaves that are young and tender does just as well. The string with which this is tied up must be well out of reach. Sweet hay should be always with them. Concentrates can be started about three weeks and, although kids are very wasteful, whatever is fed to milkers must be fed to them, otherwise they will not eat it when adult. Kids must have a salt lick within reach; a pinch of minerals should be added to bottles at a fortnight old, and to concentrates as soon as they are eating well. Soaked beet pulp can be given after they are four months old, but not before, because it is quite indigestible.

All goats, whether male or female, need their feet to be cut every month and should be dusted with louse powder every three months. With zero-grazed herds it is not necessary to dose continually for internal parasites. Once a year is adequate, unless they come in contact with other goats—at shows, say—when more often is advisable. Fed this way, six milkers, two kids and two males can be kept on $1\frac{1}{2}$ acres (0·6 ha) of land. The land must be in good heart, however. Of this, 1 acre (0·4 ha) should be herbal lea and the $\frac{1}{2}$ acre (0·2 ha) used as plough ground: this should include lucerne to be cut green—it is a semi-permanent plant and will stay for four or five years. Roots and kale, winter rye and winter oats should also be grown on the $\frac{1}{2}$ acre. The land must be heavily manured or composted for kale, which is first in rotation; this is followed by roots, succeeded by the rye and oats. The lucerne plot should be heavily manured after the last cut. The herbal lea should be given a thick coating of slurry from any cattle farmer near by; most farmers are delighted to supply this because they invariably have a surplus for which there is no demand. Ten animals do not produce enough manure to fertilize the lea.

In spring, the lea and lucerne are chain-harrowed, then rolled to flatten molehills. Weather permitting, you can take a first cut of rye about the third week in February; this is just when kale is finishing so, with careful management, you can make sure some sort of greenstuff is available all the year round. Rye is followed by oats, then on to the lea, with lucerne last.

You can now stop feeding hay, except when the greenstuff is wet, or at least cut down to one night's feed. The moment rye and oats are finished, the ground is heavily manured and kale sown: we have found 'Canson' kale the best and hardiest. Marrow-stem is liked by the goats but a very heavy frost in November can destroy the lot. Roots are sown in the middle of April, and the kale is best sown in two lots—just in case, in a dry season, thinnings are needed in July or August to keep up milk yields. All kale and roots must be kept well hoed; lucerne also hates weeds and will die back if allowed to get choked. You will have been cutting some part of the lea for greenstuff; this part will grow about three times during the season and give two cuts of hay, providing all the hay needed for ten animals, with a little over. A strip of comfrey is a great standby during a wet season, as it is possible to shake it nearly dry; it yields about five cuts—the same as lucerne.

Breeds and Breeding

To get the best return out of this system, you want big, placid, deep-bodied goats, to give a lot of milk in long lactations. Nothing must be too good for them, or too much trouble. British Saanens or Saanens seem to fit best, since they are content in yards or boxes. Many breeders wait until the kids are about eighteen months old before putting them in kid, but this is a waste of time and money. As long as you decide early, at the kid's birth, in fact, to mate her as a kid and look after her very well, nothing will go wrong. It must be realized that your kid has got to be big, strong and fit; if so, it will be safe to mate her about seven to nine months old. After mating, feed her extra well, remembering that she has to keep growing as well as feeding two or three kids. Unless well cared for, she will never make a good heavy milker and you will have wasted time and money.

It is, of course, better to breed your own stock, but one has to start somewhere. The best course for the beginner is to make an appointment to see an old-established breeder and take his or her advice. Buying good stock at the start is more expensive, but will never be regretted. Good goats are not cheap: many years of careful selective breeding and exhibiting produce the best. A good goat with a long recorded line behind her could well cost £100 or more. However, you might be lucky enough to find one with a 'show fault'—say a few coloured hairs on a white goat—and this, put to a really well-bred, milky male could breed useful animals.

Some goats will live and breed for about twelve years, while others have an accident early in life, so buying a young goat is a good proposition. As a rule, what you want at the start is a good milker. Many breeders are overstocked in spring, when most goats kid, and some will sell cheaply to a good permanent home.

By whatever method you finally decide to keep your goats, try to go to a breeder using the same system as the one you intend to adopt. It is no use taking a goat from a zero-grazed herd and putting her out into a lovely field filled with lush grass: she has no idea what to do, so stands at the gate calling to be let in. The idea of eating 'off the floor' will not occur to her and, even if it did, she would not eat enough to give you any decent quantity of milk. It is, however,

possible to transfer originally free-range goats to a zero-grazed herd
with success.

PHILIPPA AWDRY

GOAT FARM NEAR EDINBURGH

This is an illustration of the problems encountered in starting a goat
farm. I decided to go into commercial goat farming largely as a
result of reading an earlier edition of this book in which David
Mackenzie argued so prophetically the case for the development of
the dairy goat industry. In fact, we are still a long, long way from
reaching the scale of operation that he envisaged, but it is true to say
that the prospects for goats are much better now than when he
wrote his original words. It is only in the last few years that the
interest in goats as a serious commercial proposition has really
gathered momentum in this country, apart from the temporary
booms created by milk rationing during the World Wars. Now there
are several factors which seem to offer longer-term security to the
would-be goat farmer: the 'self-sufficiency' movement; the general
interest in natural foods and, in particular, the opening of 'whole
food' shops; and the gradual recognition by the public of the
benefits of goats' milk in the treatment of various complaints. This
last, although, in the eyes of the goat's advocates, still pathetically
meagre, is certainly the most significant in the long run.

In spite of this, the greatest difficulty the goat farmer encounters
is the question of marketing. I only got my operation off the ground
at all when I moved from Cumbria to a farm near Edinburgh. The
former site offered a brisk, but seasonal, trade with the tourists, but
even this depended on the allure of farm-gate sales. Efforts to find
retail outlets in the larger Lake District tourist centres met with
total failure: attempts to foster an all-year-round local market in the
small towns of the West Cumbrian coast brought only a very limited
success. At the end of a few years it became obvious that the only
way to earn a living from goats was either to mount a massive
transport operation or, better, to move closer to the larger centres
of population.

So it is only since I came to my present farm, fifteen miles from
Edinburgh, that real progress has been made. In many ways,
indeed, Edinburgh seemed to be an ideal situation with its high

proportion of academics and professional men, plus the fact that there was no large goat herd in the neighbourhood. The latter fact has proved a dubious advantage, as it has meant that the population has no communal experience of goats' milk; no stories of 'wonder cures' for eczema sufferers are part of the local folklore. On the whole we have had to depend on people's natural curiosity, and time: time for the word to get round; time for Mrs A to tell Mrs B. what the milk did for her baby; time for Mr X to tell Mr Y about the incomparable flavour of the cheese; time for just a tiny fraction of the population to accustom themselves to the idea that they could buy goats' milk, yoghourt and cheese in their local 'whole food' shop or health store and would actually enjoy it if they did.

There are other reasons why the aspiring goat farmer must allow time for his operation to become viable and why it is a project that should be worked towards over a period. One cannot go out and buy a large herd of goats and establish them overnight. Those who have tried to buy one milking female will have some idea of the impossibility of purchasing a herd of fifty. This can only be done by seizing every opportunity that comes—and those that come will be very few and will have to be sought very assiduously—and by having a carefully thought-out breeding programme. For goat farming to offer any prospect of success, only good—or preferably, very good—stock must be entertained, and the sad fact is that very few goats of either category ever become available for purchase on the open market. The only hope of building up a herd of sufficient standard will be to penetrate the 'old boy' network where most of the transactions involving the right sort of animal are carried out. Here, at least, the novice will have one vital factor in his favour and that is the friendliness and approachability of the vast majority of established breeders.

Three other factors are all-important. One is luck. I bought my first goat unseen from a dealer advertising in a national newspaper. I might have got anything: what I did get was an animal which has given me 300 to 350 gallons (1,364 to 1,591 litres) of milk every year since and transmits this characteristic to her daughters. She is unregistered, the only horned female I have ever owned and the only one I ever will own, but she is, apart from that, the ideal commercial goat.

Another thorny problem encountered in developing a suitable herd is the choice of a male. Here the opportunities are far greater because a great number of stud males regularly come on the market. Most of them should never have reached this highly privileged position and are probably doing as much as anything else to hinder the progress of the breed; but there are still some very good males available—and the male, as is well known, is half of the herd.

The last factor is the matter of culling. The breeder for the show ring has long known the importance of this, but it is equally important for the commercial farmer. The profits of goat farming will at best be so slender that passengers cannot be carried. The puny kid and the disappointing milker are luxuries that cannot be afforded. But the commercial farmer has one advantage over his show-ring orientated colleague, who must destroy the animal that is not up to the standard of his herd. The commercial man has an animal which may well be ideal as a family milker, and can be sold as such with a clear conscience.

The fact that the market will be comparatively slow to build up is something of which the goat farmer must necessarily take account because his herd will build up slowly too. We have been working for the last five years to build up a herd of fifty goats, each producing at least 250 gallons (1,136 litres) a year—or, better still, 450 to 500 gallons (2,045 to 2,273 litres) in an extended lactation. In another five years, if we are very lucky, we may achieve this target; by then, if not before, the market will need another goat farmer in this area and he will certainly be an ally, not a rival.

Obviously there are many other problems attached to such an infant industry. Finding a suitable farm is a major difficulty. Although there is some flexibility here, it is at least highly desirable that the farm should be capable of feeding the stock, even if the concentrates are to be brought in. This means growing fodder crops to cover most of the year and producing really good hay; this last is so important to the well-being of the whole enterprise that to rely on the fluctuations of the market, and the availability of the right quality, is to invite disaster. Unfortunately, one is still at the mercy of the weather.

Having procured the right farm, with a suitable steading, and about a ½ acre (0·2 ha) of good, productive land for each milker, and having budgeted for its purchase or rental, then remember the 'incidentals'. You will need a milking parlour, washing-up room and dairy that will enable you to look the local Milk Officer in the eye; you will need dairy equipment and a milking machine; you will probably need to adapt the housing, build stalls, etc., and to re-fence the land, either to make better use of its resources or to render it goat-proof; you will need a tractor and trailer, at least, and that will leave you at the mercy of contractors; you will need a van; you will need to allow for at least a year or two of running at a loss—in short, you will need the best part of £10,000, once you have bought your farm or budgeted for its rent. And no bank is going to rush very enthusiastically to your aid.

All this might seem as though this section is addressed solely to the very rich or the blindly optimistic. But this need not be so. Provided there is one substantial wage-earner in the family, ways and means might well be found, provided the final goat is held to be worth the hard work and sacrifice. The most successful goat farm in the world was started by a Californian businessman who was rightly annoyed by the difficulty of obtaining goats' milk for his eczema-afflicted daughter. But if goat farming is really to develop in this country, it will probably find its best recruits from within the farming industry. Many farms could be profitably developed to include goats as an important sideline: the principals would have the great advantage of having an experience of the basics of general agriculture and, if they were lucky, of dairy procedure too.

But they—and anyone else venturing into this field—will inevitably find that they have much to learn, simply because they are trying to turn into a commercial proposition an animal which has in this country been basically the province of the hobbyist. This is not to belittle the work of the breeders who have made the British goat the finest in the world, but it is to say that many of the methods they have used are not applicable to the commercial farmer. Kid rearing is a good example. The initial problem is of how to feed them; bottle feeding is obviously impracticable with the numbers involved, and we have tried various forms of multiple suckling, none of which seem satisfactory because of the impossibility of knowing just how much milk each individual is taking; we are now trying pail feeding, which at the moment seems to be working well, but we shall have to wait a number of years before we can really judge the long-term results.

Early weaning is also desirable, on grounds of both economy and labour, and is advocated by American writers, as well as by current practice with dairy calves. All the same, it is essential on economic grounds to be able to mate the kids their first season, which we do by retaining only the early kids as herd replacements, and by making sure that they have reached at least 100 lb (45·4 kg) weight by Christmas so they can be mated as soon as possible in the New Year. The goatling really has no place in the commercial herd, although at the moment we keep a few of the most promising of the late kids through to their second season.

Much of this flies in the face of current 'orthodoxy', although it is common practice in other countries. But, after all, we are dealing with British goats, and hoping to reap the benefits of their excellence, whether we are going to pay in the long run for the short-term advantage, only time and carefully-kept records will tell. In this, as in so many matters, the goat farmer has little objective scientific research to guide him: there is only a mass of prejudiced and highly subjective opinion.

There are matters in which the goat farmer faces peculiar difficulties. One is the lack of a stable market price for the stock he is rearing. There is a demand for kid meat in the high-class butchery trade in this area, but as yet we have not had the opportunity to test its extent and reliability. Ironically, however, the female kids present a greater problem than the males because, in spite of the heavy demand for milking goats, there is very little market for females, however well bred, unless coming from one of the very top show herds. The situation may well arise when potentially excellent stock is either being slaughtered at birth or reared for meat; certainly at the moment we find very little percentage in rearing females who are not wanted as herd replacements. It will be a disastrous situation for the future of goats in this country if good young stock is being slaughtered, while a broken-down scrub goat attracts hordes of prospective purchasers because she is nominally 'in milk'.

One possible remedy may come when the price paid for a good milking goat more nearly represents her value as a producer, and the cost of rearing her. The goat owner who spends two years lavishing expensive care and attention on his animal and then sells her for £50 or less (in 1977), is doing a grave disservice to goats in this country as well as chalking up a fairly heavy personal loss. In this, as in other matters, the sooner the goat industry is exposed more to harsh economic reality and less to sentiment, the better it will be for goats and the quicker they will achieve their proper role in British agriculture; and in this the self-confessedly commercial orientation of the goat farmer can play a vital role.

There are other matters which are less controversial, but where the goat farmer finds himself without reliable guidelines. Although one or two of the leading animal-feedstuff compounders are producing a concentrate aimed specifically at goats, there is still a great deal to be learnt on this subject. Even more important is a study of the desirable size of the concentrate ration and its effect on milk yield. Goat owners generally have not been orientated towards discovering what is the smallest ration permissible to allow the goat to achieve maximum efficiency; but this is a vital question to the farmer. The housing of the commercial herd also raises the problem of whether the well-being of the animal depends on the initial expense in materials, and continuing expense in labour, of an individual stall. Our present system is to house most of the milkers individually; but we should be the first to admit how inefficient this system is, and as our herd becomes more and more home-reared and therefore, we hope, more compatible, we shall switch to more communal housing.

The British goat is the best in the world. She has been made so by devoted breeders, most of whom have been able to give financial consideration second place. Probably much of what is written above will strike them as quite unworthy of the animal that they know. But the challenge facing the commercial farmer is to adapt this somewhat pampered animal into the useful economic unit she undoubtedly is without losing her good characteristics. The commercial goat will probably never achieve the startling records of some of the great yielders of the past, but a consistent rate of production at a much lower level than this can be profitable within the qualifications already specified. There does seem scope for the development of the dairy goat industry, which can make useful, if minor, contributions to the health and economy of the country, provide the farmer and his family with a modest living, and install the goat in her rightful place in twentieth-century Britain—as a useful producer of a beneficial food.

ROBERT HASLAM

MARGINAL FARM IN WIGTOWNSHIRE

This example shows how a few goats can be usefully integrated into the workings of a small mixed farm. The farm, with an area of 60 acres (24 ha), is in a high rainfall area, with very poor soil—sand on rock with an area of peaty bog, plus large patches of gorse (whins in Scotland) full of rabbit burrows. The only trees are hawthorns and the situation is extremely exposed to the prevailing south-west wind. Dry-stone walls surrounding the farm and dividing fields afford some shelter to sheep and out-wintered Galloway cattle. The original excellent solid stone farm steading included a large byre used for storing hay, and a goat house with four boxes roughly 5 ft 6 in (1·7 m) square and one 5 ft 6 in by 10 ft (1·7 by 3 m) for kids. In one corner were the milking bench and scales and, leading out of the goat house, the dairy area containing a large sink with hot water from an electric geyser, and a worktop to take the separator and other equipment. Half of a two-stall stable adjoining became the food store, the second stall becoming a general farm emergency room. The building was extended with passages and small yards to hold two large farrowing pigs, while the male goat house, with a large exercise yard, was established within the old stock yard. From here the billy sees the comings and goings of people and animals; he is kept in with a 7-ft (2·1 m) wire fence with posts concreted in to control both goat and pig.

A 12-ft × 10-ft (3·6 × 3 m) shed with half door and skylight was built of rough forestry slabs, the outer ones set vertically, the inner horizontally, with straw packed tightly between. Though this was originally intended for sows and litters, it became the kid house. Outcrops of rock make a splendid playground for baby kids; an area is fenced off with 1½-in (3·8 cm) mesh netting with one electric wire above. Kids go to this house and run after leaving their dams at four days and stay here for two weeks before rejoining the herd. After this interval, they are immediately recognized by their dams, but no attempt is made to suck. Kids return to their own quarters when the goats are housed. In reasonable weather they are only shut in last thing at night and can be seen playing the wildest games till late evening. It is autumn before they go to live in the goat house.

Ducks (Khaki Campbells) have been introduced and a duck house has been built near the bog: ducks are believed to eat liver fluke. The loss of a ewe from fluke prompted the idea, but whether or not the ducks have been the reason for no further case in either sheep or goats one can only speculate.

The menace from foxes, prowling round early in the morning waiting to snatch a lamb while its twin was being born, was overcome by keeping the ewes in a field near the house until the lambs were a week old.

After a soil analysis, slag and ground limestone were applied and each year 2 or 3 acres (0·8 or 1·2 ha) is sown to oats, undersown with 'Clifton Park' lea. The ground is cleared by sows—folded by an electric fence—which energetically tear out bracken and other weeds, fertilizing the ground as they go. We have had great crops by using this method, grazing the first year after harvest and subsequently cutting for hay.

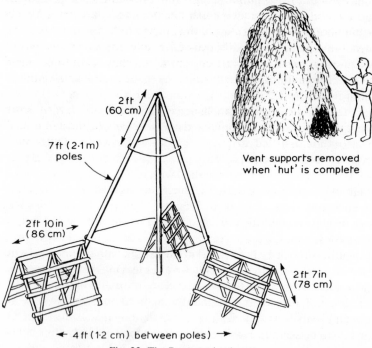

2 ft
(60 cm)

7 ft (2·1 m)
poles

2 ft 10 in
(86 cm)

2 ft 7 in
(78 cm)

Vent supports removed
when 'hut' is complete

◄ 4 ft (1·2 cm) between poles) ►

Fig. 22. The Proctor tripod system

The Proctor tripod system, far superior to the usual fixed tripod, is used for harvesting both hay and oats. Hay is built on wires and air vents; the latter are then pulled out to leave an almost vertical hut-shaped rack with the hay hanging just clear of the ground. Air circulates in the wide space inside and rain is easily shed by the steep sides. Since you cannot rely on perfect weather for the baler to pick up hay in mint condition three days after cutting, tripods are the answer—and there is no comparison between the Proctor tripod and any other system for obtaining well-got hay, which is so important to the dairy goat. Small areas can be cut as convenient for the requirements of the small-scale goatkeeper and quickly tripoded. The deep-rooted leas defy drought and the hay ensures a supply of minerals and trace elements to grazing stock. The goats have always milked very well on this diet.

Electric fencing is used to protect the hay and corn ricks while the goats make the most of the aftermath. June hay-making allows the re-growth of fresh high-protein grass to offset the seasonal drop in protein and in milk; and a fresh boost comes after harvest with September weather and aftermath grazing producing milk yields as high as at any time. The absence of flies and hot sun at this season are much appreciated by grazing stock and they seem to get their heads down and instinctively stoke up against the lean months to follow.

Potatoes are grown; the undersized 'chats' are fed to sows together with separated milk and a small dash of meal. Cooked potatoes are also fed to goats at mid-day in winter. Other feed crops are kale ('Canson' is recommended as the best), carrots, fodder beet, and cabbages for winter keep.

Goats are the first on to any new lea, followed by sheep and then by Galloway cows and calves. This system breaks the worm cycle and, once it is established that droppings tests have produced negative worm counts every time, the goats have a linseed-and-turpentine drench in the spring and garlic tablets at the rate of about two twice a week, raised to a daily dose for two or three weeks every spring and summer. Ewes and lambs have a worm drench in spring.

Stocking on the farm is at the rate of three milking goats, with probably two followers; ewes started at ten and worked up to forty-five on the improved grazings, and there are five or six Galloway cows with calves sold off at the autumn calf sales. A bull is hired for six weeks to get them in calf. There are four breeding sows—pedigree large Whites and two litters annually—and a dozen laying hens (R.I.R.xL.S.). Apart from the ducks, the picture is completed by geese: two and a gander (but they have nearly always had nests robbed by foxes).

Numerous adders sun themselves on rocks: goats have never been bitten, dogs escaped by some miracle, and the only (non-fatal) casualty so far has been one of the farm cats. Permanganate of potash, always at the ready, is effective if applied to a bite within a short time. On another farm a dog, bitten in the throat, was treated with dry crystals followed by a solution and it suffered no ill effects. Another, treated after a time lag of at least twenty or thirty minutes, with solution only, was all right and there was no swelling or subsequent discomfort. Unless provoked, an adder slips away, which is probably the reason why stock are seldom bitten.

In conclusion, the goats, thanks to clean ground and the advantage of deep-rooted leas on ground free of chemicals, herbicides and pesticides, have enjoyed complete freedom from disease without resort to drugs and vaccines. Concentrate feeding, even for goats giving $1\frac{3}{4}$ gallons (7·9 litres) plus, has never exceeded $2\frac{1}{2}$ lb (1·1 kg) daily, with no high-protein cake. Seaweed meal and iodized salt licks were the only mineral supplements.

Goat's milk is invaluable on a farm to other stock: it puts a bloom on weaner pigs, which enhances their value, while lambs, calves and foals thrive on the milk, which is suitable for all stock and is far safer and more digestible than cows' milk. The household benefits from all the derivatives: butter is easily made in a bowl with an egg whisk, a most palatable cottage cheese involves the minimum of labour, and yoghourt needs no other equipment than a vacuum flask (see p. 279). The inclusion of goats into our scheme has worked entirely to our advantage—once it was accepted that twice a day for 365 days of the year, the goats must be milked and tended.

JEAN LAING

Chapter Seventeen

The Export Trade

The prosperity of the export trade in pedigree dairy goats rests on the success and development of goat-stock improvement schemes in other, and mainly hotter, countries. Whether the schemes be supported by private enterprise or government policy is immaterial; it is the prime interest of export breeders to ensure that the schemes are successful. The quality of stock under British conditions is no indication as to their value in a different economic and physical climate; the notion that the customer knows what he wants is antiquated in other fields of commerce, and highly optimistic in this one.

By far the greater part of Britain's export trade in pedigree livestock is concerned with animals whose inherent merit is visible to the naked eye. Beef and mutton conformation, and the lines of a thoroughbred or draught horse, are bought and bred by eye. The productivity of the dairy cow or goat is a more elusive factor which cannot always be exported, and may be valueless in a new environment. Many factors are involved in the inheritance of high milk yield, but they may be conveniently regarded in two groups: those which effect food-intake capacity, and those which affect food-conversion efficiency.

The hungrier half of the world is becoming somewhat weary of receiving imports of high food-conversion efficiency in the form of 'improved' animal and vegetable varieties. Yet they continue to accept it. It remains acceptable, as rhubarb is acceptable to poor relations, not for its value, but to avoid offending the donor. In fact, 'hybrid' maize and 'hybrid' poultry have utterly failed to raise the standard of living in agriculturally primitive countries, for hungry lands cannot provide either with the nutritional ingredients they need to show their 'hybrid' virtues. Marie Antoinette was a well-intentioned and sound dietician who prescribed cake for those who cried for bread. She has a distinguished modern following, to share her fame in the future.

Though it saves human labour, high food-conversion efficiency is always expensive to maintain. Neither plant nor animal can concentrate a great deal of energy in its products unless the energy is readily available to it. The further afield roots, mouths, or beaks must forage for nutrients, the more of the food is burnt up in the search. Good feeding starts in the soil; but if we make ample nutrients available there, by lifting and carrying them from elsewhere, we set ourselves in direct opposition to the forces of soil chemistry, bacteria and climate which work to reduce nutrients to a normal and natural level; the job is an expensive one. To support universally an English diet we need to dump annually on the door-step of every household in the world the equivalent of 1 ton (1·02 tonnes) of minerals, including 5 cwt (254 kg) of coal. It cannot be done. High food-conversion efficiency in crop and stock must remain of very limited value in a great part of the world.

The good forager, animal or vegetable, which ranges far for dilute nutrients, and does a little more than feed the labourer who grows it, remains the most valuable type of stock and crop in most of the world. Of all the products of scientific plant and animal breeding, the appetite of the modern dairy goat must rank exceedingly high as a universally helpful export. The top yields of the modern dairy goat require a standard of feeding which the earth in general cannot bear; in the occasional combination of high living standards, high labour costs and dry climates, the high-yielding goat has an immediate future—e.g. in the developing dry zones of Australia, the dry zones of the U.S.A. and in parts of Africa. But, for the most part, it is the modern goat's appetite which is required to sustain currently common yields on cheaper fodder.

There is no doubt that the appetite of the modern dairy goat can be exported; indeed, the limited and uncoordinated evidence available suggests that under warmer conditions the goat's appetite for suitable fodder acually expands; but to use that appetite to best advantage, and preserve it in future generations, requires special care.

The appetite of an individual goat is highly variable. Her appetite, when in full milk, for fodder of a suitable and palatable kind, will normally be at least twice as great as her appetite when dry; and in extreme cases may be eight times as great. Given the breeding for high yields on bulky foods, the goat's appetite will vary with stage of lactation and the stimulation to milk production that her diet provides. In other words, for maximum appetite, and therefore for maximum economy in the use of fodder and land, the diet must be balanced for milk production, and must either comprise or be supplemented by a large fluid intake.

Under British conditions, the water-to-dry-matter ration for maximum yield is 4 to 5 parts water to 1 part dry matter by weight. Under hot and dry conditions, it will be wider. Then it may prove desirable to supplement a more or less desiccated bulk diet by concentrates in gruel form.

Because relative capacity for fodder has never, as yet, been a distinct and conscious objective in goat breeding, the characteristic can, in fact, only be imported in conjunction with high individual yield and in the form of rather large goats. On poor pastures, these big goats cannot move or eat fast enough to do justice to their appetites; in soils minerally impoverished by heavy rainfall, the mineral content of the herbage may be too low to maintain the massive bone needed to support a large body and yield.

Size can be reduced by outcrossing to smaller native breeds, or by arranging quickly successive pregnancies for future generations of imported goats. Relative food capacity has no positive relationship to size within certain limits. Goats weighing about 80 and 170 lb (36 and 77 kg) may have very similar food capacity in proportion to their bodyweight. Goats weighing over 170 lb often have exceptionally large appetites, the usefulness of which is limited by the ability of the goat to satisfy it on normal pastures. Goats weighing under 80 lb, when adult, tend to have relatively smaller appetites. On poor pastures, the herd of smaller goats has the advantage of the herd of larger ones in possessing more legs and mouths to gather the requisite fodder; the efficiency with which they convert the fodder to milk is substantially the same. In hot lands, the smaller goat, with a relatively larger surface area for evaporation, can keep cooler.

But no good purpose is served by breeding goats any smaller than the pasture and climate decree; it is more expensive of labour and housing to extract a gallon of milk from three goats than from two. Moreover, at below 80 lb (36 kg) bodyweight, appetite tends to decline. In all warm climates, the high ovulation rate of the goat makes pregnancy more expensive of the goat's resources, of minerals especially; in tropical countries, two breeding season a year, or an all-year-round breeding season, make it virtually impossible to maintain size within the ideal range without restricting pregnancies. The bigger the goat and yield per head that is required, the more severely must pregnancies be restricted. Biennial breeding is necessary to maintain individual yields comparable to those obtained in Britain.

With the best of care, it is probable that the yield and appetite of the goat will decline after a few generations in a hot climate, unless breeding and rearing under cool conditions can be arranged. Cold is an important factor in the development of food capacity. Cold conditions during the first few weeks of life compel the heart to produce a stronger pulse to maintain body temperature, and result in the development of more widely sprung ribs and more space in the body cavity. Thyroid-gland development, which has a profound effect on both appetite and yield, is also stimulated by cool conditions of rearing. Since fibrous foods constitute the goat's main source of heat, the consumption of such foods, and the development of digestive organs capable of accommodating large quantities, can only attain their maximum in a cool climate. In hot countries, the pedigree-breeding, kid-rearing and stock-improvement centres should be sited in the mountains.

A special difficulty is presented by the existence in some warm countries of a native long-haired goat stock. It is clear that there is the minimum of encouragement to bulk feeding if the massive fermentation resulting therefrom takes place under a blanket of hair and a sweltering sun. If it is desired to 'improve' such a stock—in the sense of sustaining yields on cheaper forage—it seems preferable to do it in two stages: in the first place, by introducing *Capra aegagrus* blood from hot countries—e.g. Jumna Pari or Zariby goats—by which the appetite of the stock will be increased and the long-haired characteristic displaced. Subsequently, the highly developed appetite of the modern dairy goat can be imported from Britain and America in circumstances which are less likely to suppress inherited characteristics by accidents of environment.

In fact, where active goat-stock improvement is in progress in the Mediterranean areas and the Near East, where the long-haired goat predominates, the aim seems to be rather to raise the average yield per head than to sustain yields on cheaper fodder. Industrialization and rising wages in Israel and Malta suggest such a labour-saving objective. Since the native goats of Malta are capable of giving their bodyweight in milk in about twenty days, under normal management, no striking improvement in their food-conversion efficiency can be expected from outcrossing; greater size and equivalent pro-ductivity may be imported, and this results in higher yields per head. But controlled selection of breeding stock and restricted breeding practice might achieve the same result. In Israel, where social and geographical conditions have been less favourable to selective breeding, the native goat stock is more variable than Malta; but still contains individuals capable of comparable performance, and the same considerations apply.

The Malaga goat of Spain, the Mamber goat of the Near East and the Maltese goat provide long-established local answers to a com-mon problem. In dry and rocky country, where natural water supply limits cultivation to patches, the produce of the patches is precious, and that of the intervening desert is negligible. These goats have a relatively low food capacity, but are highly efficient in converting concentrates (beans and barley) to milk. In the more arid and rocky regions, a goat with a bigger appetite might have to range so far to suffice it with uncultivated roughages that efficiency in converting concentrates to milk would be impaired. If more milk is needed from these acres, the face of the land must be modified as well as the goat.

Economy in milk production is not an easy subject to discuss in an international way. It is a most confusing subject to consider even on a local basis. There is an obvious appeal in the notion that a lot of milk from a few goats is more cheaply produced than the same quantity from many; under such circumstances it may be true. But, even on the highly cultivated fields of England, where wages are relatively high, the economic statistics suggest that profitability is more closely related to yield per acre (0·4 ha) than to yield per head. Where cultivation is less intensive and labour is cheaper, this ten-dency must be more marked. Production per unit area of cultivable ground is probably the soundest basis on which to estimate dairy economy in any land. Only when production is sustained entirely by the 'spontaneous offerings of nature' is yield per head a sure criterion.

Though the appetite and short coat of the Swiss breeds are a helpful export to most parts of the world, other of their breed characteristics are not. The prick ears of the northern goat stock have, in their homeland, an advantage over the lop ears which predominate in hot countries, in that they are more easily warmed extremities and less expensive of the goat's heat resources in a chill wind. But lop ears perform such a variety of functions in a hot climate that it is a significant handicap for a goat to be without them. They protect the goat's face from sun glare, insects and blown sand; their large evaporation area and constant, effortless movement cool the goat; and the ragged contours of the lop ears of most goats of the East evidence the protection that these relatively insensitive appendages provide against thorns and dogs and similar misfortune. Most of the cool contentment that midday sees between Valetta and Rangoon is in the swaying shadows on the face of a cudding goat.

The wedge-shaped profile is inseparable from big appetite and useful productivity; but if the wedge has the horizontal top line of the Swiss breeds, the udder is carried too near the ground for safety among desert thorn. The sloping top line of the desert goat (Fig. 14, p. 186) can support a wedge-shaped profile and still insure the goat's udder against the hazards which prevail everywhere but on cultivated fields and alpine pastures. A big goat of Swiss type may appear to carry her udder high and safe on long legs; but it is still the lowest point of her under line, and takes the whack as surely as the sump of a Jeep.

For many years British pedigree breeders attempted to fix a breed type which combined the most useful features of the goats of the East with Swiss appetite and productivity. These breeders pursued their purpose in the light of an unmatchable tradition of stock breeding, and practically unimpeded by immediate commercial considerations. In the modern Anglo-Nubian, they have achieved a fair level of uniformity, and as near an approach to their ideal as could be hoped for under the circumstances. Yields expressed in pounds of milk do not compare with those of the Swiss breeds; but expressed in pounds of milk solids they do; at the price of milk, water is dear in any land. Lop ears are well established; the sloping top line has been neglected and even penalized in the show ring, but persists in many strains. While I have no vested interest in Anglo-Nubians, it appears to me that the man who imports modern dairy goats of Swiss type into a hot climate will be outstandingly fortunate if he breeds therefrom in his own lifetime a single goat so well adapted to his needs as the average Anglo-Nubian is ready-made to be.

That is an opinion, honest even if not acceptable. What is needed at both ends of an export business is knowledge. The breeder for export requires to know how his stock perform under foreign conditions, in what way they excel and fail; the importer requires to know more about the stock he buys, especially about those inherent characteristics which are of small importance in this country but great importance in others—fertility, growth rate of kids, etc. Neither of these requirements can be met without controlled recording at both ends and an interchange of views. Tight organization of breeding at both ends of an export trade is an essential to success.

It is more than probable that the accentuation of qualities required by overseas goat farmers may make the British goat and

Plate 18.   R124 Berkham Butterlady, Q*7 BR. CH. Miss E. Rochford's very elegant Anglo-Nubian.
(Photograph: Diane Pearce)

the British Goat Society more interesting to farmers in this country too.

The export trade is exceedingly brisk. During 1976, ninety-two goats were exported through the British Goat Society to Nigeria, Trinidad, Jamaica, Eire, Northern Ireland and Canada. These were mostly Anglo-Nubians, Saanen and British Saanen, with fewer British Alpine, Toggenburg and British Toggenburg. In 1977, enquiries came again from Eire (British Alpine) and from Nigeria (British Saanen) and additionally from St Vincent (Anglo-Nubian), India (Saanen and British Saanen), Brazil (Anglo-Nubian) and Thailand (British Alpine and British Saanen). Since its tourist trade has been disturbed by the recent political upheaval, Portugal has shown renewed interest. Previous years have seen goats going to Bermuda, France, St Helena, Barbados and Dominica. Private sales include a batch of sixty to Sudan (Anglo-Nubian, Saanen and British Saanen) in 1976; previously, goats of various breeds have been sold privately to France, South Africa, Kenya, Japan and Malta; and lately Anglo-Nubians to Oman, and others to Yemen.

In 1979 about 250 goats were exported from Great Britain, but the main consideration for 1980 was felt to be price. West Germany is now one of the main exporters of goats and British exporters have found that the high cost of British exports has made them less than competitive, despite the fact that goats all over the world have become immensely popular, particularly in developing countries, which have come to realize the value of the goat over the cow. Export prospects were, however, felt to be good at the time of writing.

Appendix One

Some Notes for Novices
by Jean Laing

General Health and Safety

On entering the goathouse a first-class herdsman has a quick look round and is able to spot immediately any animal off-colour. The ability to do this is a gift beyond price. Serious illness may be averted by First Aid. A really sick goat is unmistakable; she stands, usually in a corner, looking tucked up and thoroughly dejected, hair possibly on end, ears drooping, extremities often cold. She may never reach this stage if her owner is vigilant. In a case of slight indigestion she could appear all right through probably not cudding. A goat invariably cuds during milking; if she fails to do so all is not well. The novice should note these signs. Goats are great animals for routine and each follows its own particular pattern; any deviation, however slight, from normal behaviour may indicate trouble brewing. It pays to be observant.

It is astonishing that some goatkeepers do not realize the necessity to take a goat's temperature and rug warmly at the first sign of illness. Taken in the rectum the normal temperature is $102\frac{1}{2}°-103°F$ ($39\cdot2-39\cdot4$ C). Moisten the thermometer in lukewarm water, shake down to 95°F (35 C) hold in the hand to keep it warm. It can then be slipped in easily and unnoticeably to half its length. Leave in for 3 minutes. After reading clean it at once and stand it in some disinfectant. The novice should seek veterinary advice if there is a rise of two degrees or slightly less. Make it a rule that any sick goat is rugged at once. Ideally the rug is made of melton cloth or similar thick woolly material: fold over, shape for neck and stitch. See the rug comes well down over the flanks and under the stomach, tie in two places, under the chest and the stomach. Kids can be rugged quickly with an old woollen jersey, adapted to fit with holes cut for the legs. In an emergency an old blanket, preferably of double thickness, will serve.

Table 9 on p. 251 shows the contents of a first aid kit for the goat house.

Withhold all food if a goat is ill. Allow access to warm water. (Garlic enemas are a great help for a serious digestive upset). If the digestion is at rest the system is able to clear whatever is causing the trouble much more easily. Warm rugging is essential and as much rest as possible. A sick or chilled goat may stand for hours unless induced to lie down. Massage limbs. If the goat has confidence in its attendant he can fold back a foreleg, persuade the goat to kneel, then to lie. Warm up with hot bottles and pack round with straw. The goat relaxes and half the battle is won. Warm salty water should be offered frequently—this helps to get rid of any poisons in the system. Do not isolate a sick goat so she cannot see or hear her companions.

CHANGES IN FEEDING
These can cause serious digestive troubles: such changes should be carried out very gradually. The often fatal disease, enterotoxaemia, can result if this precaution is not heeded.

EXTERNAL PARASITES
Lice probably cause more loss of condition and milk than any other factor. The novice may be unaware of the presence of lice, on account of their small size and retiring habits. As a routine measure, dust all the goats with derris, which is the safest insecticide, then repeat in ten days' time, subsequently every month from September to two weeks before kidding. Turn back the hair from the root of the tail to the shoulder and up behind the ears, sprinkling and rubbing well in as you go. Repeat round the shoulders and behind the front legs, making sure that the powder reaches the skin.

POISONS
If a goat eats any poisonous substance, prompt action is vital and directions are set out in Chapter 10. Remember the main principle: first remove the goat from the poison, *then* the poison from the goat! Treat by giving a bumper dose of Epsom salts, 6 oz (170 g) in half a

pint of warm water. Withhold water which would swell grain and make the situation considerably worse. In severe cases send for the vet. If the food is grain he will give an intravenous injection of a Vitamin B complex. Chlorodyne may be given to alleviate pain. Treatment for shock, i.e. rugging and warm stabling, is essential, and later a mild dose of linseed oil.

RAIDING THE FOOD STORE
It is a golden rule that bins or bags containing concentrates should *never* be accessible to stock. Being exceedingly artful a goat is quite capable of pushing up the lid of a bin or even worrying away at the catch in order to do so. There should never be the slightest possibility of this happening. If it does, speedy action is vital or the goat will probably die—painfully.

OTHER ACCIDENTS
Avoid these by taking simple precautions. Never leave a collar on a goat, nor leave tools lying around. A fork leant carelessly against a wall can irrevocably damage an udder. Barbed wire is taboo for the same reason. Kids frequently hang themselves in loops of string, or break legs jumping into low hay racks. Goats slip kids and suffer serious damage by crowding in doorways. Broken legs or worse can result from slippery paths.

Dressing Wounds:
Have some cotton wool, plenty of it, and two bowls of disinfectant. Gently swab the wound, discard the used piece of wool into the second bowl, take a clean piece and repeat till the wound is clean.

Some herds suffer frequent deaths, illnesses or accidents; but owners blessed with a little common sense and know-how survive for decades without serious trouble. By evolving a routine, and sticking to it, they get maximum enjoyment and efficiency from goat keeping.

Mating

Novices sometimes find difficulty in deciding whether or not a goat is in season. Behaviour varies, but at least some of the following signs will be apparent. There may be frequent calling, or rhythmic side-to-side motion of the tail; the vulva may be enlarged, and a slight sticky discharge observed; the goat will mount, and be mounted by, her companions; she may be restless and excitable and look about wildly for a mate. The milker's yield rises before coming into season, drops for a day or so, then rises again. In heavy milkers, the effect is pronounced, and can be a useful advance warning. If there is difficulty in deciding whether or not any goat is in season, get a stud-goat owner to rub a woolly rag over the male and seal it in a plastic bag. This oderiferous object is then opened and handed round for milkers and goatlings to sniff. The reaction of any in season is unmistakable.

At the beginning of the breeding season the goat will probably accept the billy for two or three days; but from November onwards she may be off very much sooner, and it is wiser to get her mated within a few hours. An appointment should be made with the stud-goat owner, and a certificate of mating secured, without which the kids cannot be registered. It is usual for a return service to be granted should the goat 'turn', i.e. come in season again in twenty-one days. If, after mating, she does not come on again, one assumes she has 'settled', i.e. is in kid. Goats, however, are more unpredictable than any other stock and there is no certainty until the kids can be actually felt moving at a much later date. A long-lactation milker is unlikely to go dry; in fact, her yield will rise in February and March if she is not in kid. This is not an infallible sign, since some milk on whether in kid or not; but most goats have a dry period. A pregnancy test can be arranged through the vet, but results can be unreliable.

Miscellaneous Problems

GOAT RESTLESS AT MILKING
Tie the goat firmly by its collar to the wall or partition; then tie from the collar to a staple or ring placed in the wall immediately above the goat's hip bone. Never feed while milking; the goat will become restless after finishing her feed, and this leads to improper stripping out. The natural reaction during milking is to cud.

KID REFUSES RUBBER TEAT
Form a piece of butter muslin, several layers thick, into the shape of a goat's teat. Dip this in warm milk, pop it in the kid's mouth, and it

will suck. When sucking is established, cover the teat with two or three layers of muslin and the kid will continue to suck from this. Finally, discard the muslin; the kid will then take the teat.

GOAT SUCKS HERSELF

Make an udder bag, using any strong material of about flour-sack thickness, to fit comfortably when the udder is full. Attach tapes on either side to meet over the back, with a loop on each. Slip one loop into the other; then take another tape, loop it around the collar, slip it through the loop from the udder and tie the ends together. It all has to fit properly, so measure as you go. A modern teat seal plaster is more effective.

Fencing

An electric fence is ideal, but only works effectively if the goats are trained firstly by receiving at least two good shocks. Tempt them, with tasty greenstuff, to touch the wire, and do not leave until all have been shocked and understand that the fence hurts. Even then, only retreat a short distance and make sure they are not attempting to approach the wire. It is, perhaps, unnecessary to point out that leaving goats on bare pastures in a chilly wind drives them past endurance to escape by any means, however painful.

The normal fence for goats consists of three wires, the top shoulder high, the lowest about 12 in (30 cm) from the ground. The lowest wire must be disconnected when the grass grows, or it will short-circuit the fence. A single wire can be used above a wall or low barrier. Wires may be placed round crops, implements, hay and corn stacks, and used to protect growing hay. Goats can be easily moved to clean ground, and the method obviates tethering. Electric fencing is respected by cattle, horses and pigs but not, judging from past experience, by sheep.

Appendix Two

The Importance of Weeds in Goat Nutrition

Constituents of the dry matter of various pasture plants
(after Prof. F. W. Fagan, Brynmor Thomas and C. B. Fairbairn)

Plant	Crude protein %	Fibre %	Calcium (CaO) %	Phosphorus (P_2O_5) %	Sodium (Na_2O) %	Chlorine (Cl) %	Magnesia (MgO) %	Cobalt (Co) p.p.m.
Bilberry	15·70	16·07	0·75	0·77		0·14		
Burnet			1·84	0·62	0·07	0·15	1·82	0·18
Buttercup	25·06	16·76	1·48	1·44	0·89	0·25	0·34	0·25
Catsear	19·55	20·06	1·45	0·81		3·47		
Chicory	19·20	13·40	2·00	1·11	0·37	0·92	1·07	0·20
Cowslip	12·64	15·81	1·55	0·54		0·53		
Daisy (stem)	14·30	18·66	1·24	1·17		1·10		
Dandelion	19·36	11·27	1·80	0·98	0·86	2·00	0·78	
Devil's-bit	12·86	21·42	1·23	0·63		0·50		
Gorse	16·84	33·10	0·71	0·74		0·53		
Hawkbit	18·62	18·82	2·38	0·90		1·88		
Heather (*Calluna*)	8·56	21·67	0·47	0·22		0·07		
Heath rush	14·03	23·92	0·18	0·50		0·46		
Hogweed	20·07	15·50	1·72	1·01		0·82		
Knapweed	20·16	19·02	1·56	1·08		0·20		
Lady's-bedstraw	16·49	22·14	0·49	0·57		0·32		
Nettles	27·45	32·67	5·99	1·75		1·01		
Ox-eye daisy	10·80	27·88	1·06	0·74		0·53		
Plantain (broad—leaved)	18·61		3·39					
Plantain (Ribgrass)	20·25	14·84	2·32	0·75	0·62	0·40	1·01	0·20
Self-heal	10·32	18·57	1·20	0·61		0·67		
Sorrel	24·71	15·38	0·71	1·37	0·27	1·25	0·26	0·22
Shepherd's-purse	27·26	26·65	2·85	1·35				

Plant	Crude protein %	Fibre %	Calcium (CaO) %	Phosphorus (P_2O_5) %	Sodium (Na_2O) %	Chlorine (Cl) %	Magnesia (MgO) %	Cobalt (Co) p.p.m.
Sowthistle	17·76	20·83	2·10	1·12		1·70		
Speedwell	12·20	22·90	1·35	0·86		0·28		
Tormentil	10·50	27·36	1·07	1·03		0·21		
Creeping thistle	29·64		2·97	1·17				
Thistle (melancholy)	17·50	14·56	4·61	0·94				
Yarrow	19·90	24·31	1·57	0·69	0·06	0·53	0·75	0·17
Yellow rattle	23·10	20·60	1·88	2·12	0·21	1·40	0·73	
Average	16·70	20·49	1·66	0·91	0·27	0·87	0·85	0·20
Legumes:								
Trefoil	17·5	28·5	2·32	0·79	0·18	0·56	1·27	0·20
Alsike	22·0	30·0	2·79	0·76	0·06	0·42	1·04	0·17
Lucerne	17·0	30·0	3·01	0·91	0·10	0·50	0·91	0·15
Sainfoin	17·5	34·5	1·44	0·89	0·06	0·14	1·31	0·16
Clover (wild white)	23·31	23·09	3·08	0·89				
Average	19·2	29·1	2·53	0·85	0·10	0·41	1·13	0·17
Grasses:								
Crested dogstail			0·46	0·53	0·11	0·56	0·44	0·18
Yorkshire fog			0·32	0·71	0·18	0·80	0·19	0·17
Peren. ryegrass	12·2	26·0	0·65	0·59	0·19	0·51	0·35	0·15
Cocksfoot			0·59	0·59	0·17	0·31	0·36	0·14
Timothy			0·58	0·53	0·30	0·56	0·42	0·15
Meadow fescue			0·62	0·58	0·20	0·66	0·44	0·16
Average			0·54	0·59	0·19	0·57	0·37	0·16

Notes: The mineral reason why grass is an unsatisfactory food for goats, especially for young stock and milkers, is most striking. The relative poverty of the legumes in sodium and chlorine is a particularly serious matter for the goat, who excretes 50 per cent more of these elements in each gallon of milk than does the cow.

Manganese is an element of special importance for male fertility. Only very recent analyses are reliable. Meadowsweet and willowherb are relatively rich in it.

The above table includes some unpublished data from Brynmor Thomas and C. B. Fairbairn of Durham University School of Agriculture.

Appendix Three

Allergies and Goats' Milk
by Jean Laing

During recent years the public has been made aware that some ailments, such as asthma, eczema, psoriasis and migraine, are the result not so much of disease as of the body's reactions to substances that are ingested, inhaled or contacted: allergens, in fact. High on the list of products which cause allergies of one kind or another are beef products, and many sufferers have been greatly relieved, often cured, by a diet which forbids all bovine products. Eggs and fish and other foods which act as an irritant to their condition have also been eliminated from the diet of such people with successful results.

More doctors are coming to realize that a change of diet rather than a course of drugs can be beneficial in certain circumstances. This is where a goats'-milk diet can be valuable. Some doctors, reluctant to admit diet as a key factor, have ultimately changed their views after the failure of various orthodox treatments in and out of hospital, and have recommended their patients to try the goats'-milk diet.

Problems arise because individuals have allergies which differ widely, but it is now possible for tests to be made on patients clearly allergic to cows' milk to determine whether or not they can eat beef. Some sufferers from eczema, asthma, psoriasis or migraine cannot eat beef, but many can do so without ill effects, which helps to relieve the otherwise stringent diet they are forced to follow.

An intelligent mother can cope with the diet for babies and young children, whose meals are prepared in the home, but for older children and adults, whose daily routine requires them to eat out, a different approach is necessary, which needs the co-operation of the sufferer. Boys unable to play football, because shorts reveal red patches behind the knees, and teenage girls with rashes on their arms, which force them always to wear long sleeves, will be found to be willing participants in a diet designed to cure these conditions. If the incentive is there, the chances of relief are very good indeed. After the basic rule of no bovine products—and these can even extend to the gelatine which coats some drugs—there are other rules to be followed. A wide range of basic foods should be avoided—fish, eggs, oranges, bananas, tomatoes and lentils and, for a few, pips, nuts, yeast and sometimes also oatmeal. Only by experiment can the individual discover which are acceptable foods, trying out each for about ten days before judging it to be safe.

This is where a goats'-milk diet can be very helpful. The milk is not controlled by regulations governing cows' milk. Goats in this island are considered to be tuberculin and brucellosis free and are allowed by the Ministry of Agriculture to mingle freely with accredited stock, so pasteurization is considered neither wise nor necessary. Heat treatment of milk destroys certain vital elements so, when heated milk is used for young children, it is necessary to put back the lost elements.

An EEC directive has recently emerged, however, which states that there is no need for goats' milk to be pasteurized, a directive to be hailed with joy by goatkeepers.

Leaflets giving further information about this subject and specimen diets are available from the British Goat Society, goat clubs and health-food stores in Britain.

Appendix Four

Metabolizable Energy and Nutritive Value of Foods

This information is reproduced, with the kind permission of the authors, from *Modern Milk Production* (Faber) by Malcom E. Castle and Paul Watkins.

Dr Castle, of the Hannah Research Institute, Ayr, Scotland, points out that although the nutritive values in the following table are used widely for devising rations for cattle and sheep, the values should also form a sound basis for calculating rations for goats.

FOOD ENERGY

An important function of food is to supply energy to the animal, and thus the energy value of food is of vital importance. For many years the contents of the digestible nutrients in a food were used to calculate a starch equivalent value, but this system of feed evaluation and feed requirements has now been changed to a system based on metabolizable energy (ME). The ME value of a food may be determined in an animal-feeding experiment using a respiration chamber or calorimeter. The method is similar to a digestibility trial, and the faeces, urine and expired gases, which include methane, are all collected. The content of energy in the food is determined, and from this value is deducted the energy value of the faeces, urine and methane. The resulting value is the ME of the food, which is expressed in megajoules (MJ) per kg dry matter (DM), as shown in the following example, which is of grass given to sheep.

Dry-matter intake per day   =   2·0 kg
Gross energy intake   =   35·0 MJQ
   Energy in faeces   =   14 MJ
   Energy in urine   =   1 MJ
   Energy in methane   =   2 MJ
        Total   =   17·0 MJ

Metabolizable energy $= \dfrac{35·0 - 17·0}{2·0} = 9·0$ MJ per kg DM

The detailed table begins overleaf:

Food	Dry matter content (%)*	Metabolizable energy (MJ per kg DM)	Digestible crude protein (% of DM)	Digestible organic matter in DM† (%)	Calcium (% of DM)	Phosphorus (% of DM)
Grass						
Grazing, high-quality	20·0	12·1	18·5	75	0·66	0·30
Ryegrass, post-flowering	25·0	8·4	7·2	55	0·45	0·25
Green legumes						
Red clover starting to flower	19·0	10·2	13·2	65	1·76	0·29
Lucerne, early flower	24·0	8·2	13·0	54	2·10	0·40
Silage						
Grass, high-digestibility	25·0	10·2	11·6	67	0·75	0·35
Grass, low-digestibility	25·0	7·6	9·8	52	0·70	0·30
Red clover	22·0	8·8	13·5	56	1·54	0·21
Maize	21·0	10·8	7·0	65	0·37	0·32
Hay						
Grass, high-quality	85·0	9·0	5·8	61	0·40	0·25
Grass, low-quality	85·0	7·5	4·5	51	0·33	0·21
Lucerne, half-flowering	85·0	8·2	16·6	55	1·93	0·26
Dried grass and legumes						
Grass, leafy	90·0	10·6	13·6	68	0·75	0·30
Lucerne, early flower	90·0	8·7	12·8	57	2·00	0·26

Straw						
Barley, spring	86·0	7·3	0·9	49	0·34	0·09
Oat, spring	86·0	6·7	1·1	46	0·34	0·10
Roots						
Turnip	9·0	11·2	7·3	72	0·48	0·34
Swede	12·0	12·8	9·1	82	0·48	0·23
Sugar beet	23·0	13·7	3·5	87	0·16	0·19
Green crops						
Kale, marrow-stem	14·0	11·0	11·8	70	2·14	0·33
Cabbage, drumhead	11·0	10·4	10·0	66	1·36	0·27
Grain						
Barley	86·0	12·9	8·2	86	0·05	0·38
Maize	86·0	14·2	7·8	87	0·02	0·27
Oats	86·0	12·0	8·4	68	0·09	0·37
Oil cake and meals						
Coconut meal	90·0	12·7	17·4	74	0·22	0·66
Cotton cake (decorticated)	90·0	12·3	39·3	70	0·32	1·47
Groundnut						
(decorticated, extracted)	90·0	11·7	49·1	75	0·16	0·63
Palm kernel meal (extracted)	90·0	12·2	20·4	78	0·23	0·56
Soya bean meal (extracted)	90·0	12·3	45·3	79	0·23	1·02

Food	Dry matter content (%)*	Metabolizable energy (MJ per kg DM)	Digestible crude protein (% of DM)	Digestible organic matter in DM† (%)	Calcium (% of DM)	Phosphorus (% of DM)
Other foods						
Fish meal, white	90·0	11·1	63·1	68	7·93	4·37
Brewers grains, barley, fresh	22·0	10·0	14·9	59	0·50	0·59
Brewers grains, dried	90·0	10·3	14·5	60	0·32	0·78
Maize, flaked	90·0	15·0	10·6	92	–	0·29
Maize, gluten meal	90·0	14·2	33·9	85	0·04	0·14
Sugar-beet pulp, molassed	90·0	12·2	6·1	79	0·63	0·07
Beans, field	86·0	12·8	23·0	81	0·18	0·66

* To convert to g per kg, multiply by 10.
† D-value.

Content of Calcium, Phosphorus, Magnesium and Sodium in the Main Mineral Supplements

Mineral supplement	Ca (%)	P (%)	Mg (%)	Na (%)
Steamed bone	38·5	13·5	0·35	0·47
Dicalcium phosphate	23·6	17·9	–	–
Monosodium phosphate	–	17·0	–	25·7
Ground limestone	38·0	–	–	–
Common salt	–	–	–	39·0
Calcined magnesite	–	–	52·0	–

Note: the values in this Appendix are based on data in the following five publications, which include more detailed compositions of a wider range of foods for animals.

Energy allowances and feeding systems for ruminants, Technical Bulletin No. 33, 1975, HMSO, London.

Feedingstuffs evaluation unit, First Report, 1976, Rowett Research Institute, Aberdeen and DAFS, Edinburgh.

McDonald, P., Edwards, R. A. and Greenhalgh, J. F. D., *Animal Nutrition*, 1975, Longmans Group Ltd, London.

Nutrient allowances and composition of feedingstuffs for ruminants, ADAS Advisory Paper No. 11, 2nd edition, 1976, Ministry of Agriculture, Fisheries and Food, London.

Appendix Five

British Goat Society Recognized Milking Competitions 1978/9
By Bob Martin, B.G.S. Shows Secretary

Regulations relating to BGS Milking Competitions were revised in 1978 to incorporate metric weighing (approx. 4·5 kg per gallon). The *Total Points*, designed to reflect both yield and quality, remain similar to those of previous years, but a very slight reduction in *Time Points* prevents direct comparison.

Butter-fat Percentages remain unchanged, as does the 3 per cent minimum qualifying level at both milkings.

The *Best Individual Milking Achievements* (points) in each breed for 1978 and 1979 are shown opposite:

| | | 1978 | | | | 1979 | | | |
Breed		Milk (kg)	Butter-fat % a.m. p.m.	Months kidded	Total points	Milk (kg)	Butter-fat % a.m. p.m.	Months kidded	Total points
Golden Guernsey	1.	5.10	3.46–3.91	4	19.99	6.3	3.99–4.73	1	26.15
	2.	4.7	3.89–4.33	3	19.49	4.15	4.32–4.46	12+	19.77
Toggenburg	1.	6.5	3.09–3.64	4	24.79	6.3	3.58–4.00	2	24.81
	2.	6.4	3.09–3.50	2	23.80	5.4	3.43–4.24	2	21.42
	3.	5.15	4.40–4.30	1	22.55	5.15	3.28–3.78	1	19.55
Saanen	1.	7.1	4.20–4.15	–	28.66	7.85	3.10–4.15	4	30.66
	2.	7.35	3.20–4.10	3	28.64	5.6	3.70–4.65	6	23.93
	3.	5.35	5.49–6.80	2	26.68	5.55	3.53–3.53	4	21.75
British Alpine	1.	5.5	4.34–5.71	3	24.92	5.7	5.04–5.28	2	25.91
	2.	5.0	6.10–5.85	1	24.37	6.05	3.80–4.40	2	24.66
	3.	5.85	4.11–4.10	4	24.31	6.35	3.50–3.75	2	24.53
Anglo-Nubian	1.	5.85	5.07–5.68	3	27.36	5.65	6.00–6.16	1	27.76
	2.	5.0	5.85–6.87	4	25.87	5.3	5.94–6.38	1	26.24
	3.	5.0	5.42–6.89	1	24.77	6.15	4.08–4.74	3	26.11

| | 1978 | | | | 1979 | | | |
Breed	Milk (kg)	Butter-fat % a.m. p.m.	Months kidded	Total points	Milk (kg)	Butter-fat % a.m. p.m.	Months kidded	Total points
British Toggenburg								
1.	7.1	3.70–3.80	1	27.56	6.8	4.52–4.79	–	28.88
2.	7.35	3.45–3.30	1	27.30	6.5	4.46–4.39	5	28.05
3.	6.8	3.30–3.70	3	26.09	7.25	3.03–4.01	2	27.61
British Saanen								
1.	7.45	4.59–4.63	12+	34.14	10.0	3.15–3.20	2	36.41
2.	7.8	4.22–5.21	3	34.01	9.0	3.80–4.03	2	35.75
3.	8.2	3.95–4.27	1	33.08	8.35	3.85–4.64	2	34.40
Any other variety								
1.	8.6	3.82–5.24	2	36.49	8.95	3.18–3.16	1	33.26
2.	8.1	4.26–4.98	3	34.93	8.2	3.94–3.79	1	32.20
3.	8.7	3.82–4.04	2	34.62	7.65	3.51–4.06	12+	32.20

Appendix Six

The British Goat Society and Overseas Addresses

The British Goat Society is the official organization concerned with goatkeeping, breeding and registration and publishes a monthly *Journal*, as well as an annual *Herd Book* and *Year Book*. It celebrated its centenary year from October 1979 to October 1980, starting with a reception in London on 16 October 1979 and followed by a variety of events around the country in the following twelve months. The Society welcomes new members, both from home and overseas. Prospectuses and details of the services offered, together with publications, stationery, etc., and particulars of herd registration, *Herd Book* fees, milk recording fees, etc. can be obtained from the Secretary, British Goat Society, Rougham, Bury St Edmunds, Suffolk IP30 9LJ.

Subscriptions (at September 1979):

Full membership (includes *Journal, Herd Book* and *Year Book*)	£8·00
Associate membership (includes *Journal* and *Year Book*)	£5·00
Junior membership	£3·50
Family membership	£3·00
Partnership membership	£9·00
Life membership	£100·00

The following is a list of the participants of each country at the most recent International Conference in 1971.

Algeria: Jean-Édouard del Perugia (expert, F.A.O.), B.P. 32—Birmandreis, Alger VIII

Jean Maiterre (Professor),
Institut National Agronomique, El Harrach

Argentina: Dr Hugo José Olaiz (Veterinary surgeon),
Callé 36, No. 1817, La Plata, Prov. Buenos Aires

Brazil: Professor Mario Halmilton Vilela,
Faculdade de Zooteenia (P.U.C.),
Caixa Postal 143, Uruguaiana, R.S. Brazil

France: Jean-Claude le Jaouen (Ingénieur Section Caprine),
Secretaire-Géneral de la Fédération Nationale des
Eleveurs de Chèvres, I.T.O.V.I.C.
149 Rue de Bercy, 75579, Paris

Germany: Dr Christian Gall (Professor, F.A.O.),
Wartburgplatz 6, 8–München 23

Professor Martin Tegtmeyer,
23, Kiel-Wik, Steenbeket Weg 151

India: Dr Ram Sarup,
Indian Council of Agricultural Research,
Krishi Bharan, Room No. 306, New Delhi

Iran: Veterinary Professor Sattari,
Zootechnical Department, Veterinary Faculty,
Teheran University

Italy: Professor Rafaele Mazzoni,
Via Severano 5, Roma

Malaysia: C. Devendra, Senior Research Officer,
Malaysian Agricultural Research and Development
Institute,
Jalan Swettenham, Kuala Lumpur

Mexico: Professor Abrahim Agraz Garcia,
Bahia de San Cristobal 3–3, Col. Huasteca, 17, D.F.

Norway: Knut Rønningen, Første Amanuensis Dr,
Box 24, Volbekk

Portugal: Francisco Cabral Calheiros,
Estacǎo Zootechnica Nacional, Fonte Boa (San-
terem)

Fernando Vieira de Sa-Veterinaire,
Rua Vale Formose de Baixo No. 1, Lisboa

Spain: Eduardo Laguna Sanz (Veterinary surgeon),
Direccion General de Ganaderia, Ministerio de
Agricultura, Madrid

Demetrio Tejon Tejon (Veterinary surgeon),
Patronato de Biologia Animal,
Av. Puerta Hierro s/n, Madrid 3

Sweden: Erik Sjödin (agronomist),
Näs 871 00 Härnösard

Switzerland: Peter Amman, Schweiz Zentralstelle Für Klein-
viehzucht,
Belpstrasse 16, CH-3000 Berne 14

Trinidad: Dr E. Laurence Iton
(Technical Officer, Animal Production and
Research),
Ministry of Agriculture,
Government Stock Farm, St Joseph

Tunisia: Roger Moucot (F.A.O.),
4 Rue Procope, Carthage

Turkey: Dr Martin Sengonca, E.Ü. Ziraat Facultesi,
Zootekni Kurusu, Bornova-Izmir

Additional Addresses, not affiliated members of British Goat Society:

Canada: Canadian Goat Society, Canadian National Live
Stock Records, Holly Lane, Ottawa, Ontario,
Canada K1V 7P2

U.S.A.: Don Wilson (Secretary), American Dairy Goat
Association,
P.O. Box 186, Spindale, N.C. 28160

Kenya: Kenya Goat Breeders' Society,
P.O. Box 44233,
Nairobi

The following are affiliated to the British Goat Society:

Australia: Goat Breeders' Society of Australia,
Box 4317 G.P.O., Sydney 2001

Holland: Nederlandse Organisatie voor de Geitenfokkerij,
S.H. Pak, Looydijk 67, Oudmaarsseveen, Maarssen
2570

Malta: Department of Agriculture, Valetta

New Zealand: Goat Breeders' Association,
P.O. Box 294, Morrinsville

Appendix Seven

British Suppliers of Requisites for Goatkeeping

All small equipment: milking pails, disbudding irons, tethers, collars, etc.:
Fred Ritson, Goat Appliance Works (Est. 1938), Longtown, Carlisle CA6 5LA
Smallholding Supplies, Priory Road, Wells, Somerset BA5 1SY

Cartons for goats' milk:
Bowater Industrial Packaging Ltd, Princes Way, Team Valley Estate, Gateshead NE11 0UT

Cheese moulds:
W. H. Boddington & Co. Ltd, Horsmonden, Kent TN12 8AH

Electric fences:
Rossendale Electronic Fencers, Lumb, Rossendale, Lancs BB4 9NJ (until 1986)

Electric netting:
Livestock Fencing Ltd, PO Box 73, Gloucester GL3 4AF

Garlic, herbs, etc.:
Dorwest Herb Growers, Freepost, Bridport, Dorset

'Optimalin' for tanning skins:
Watkins and Doncaster the Naturalists, Four Throws, Hawkhurst, Kent

Rennet, colouring, etc.:
Chr. Hansen's Laboratories Ltd, Reading RG2 0QL

Seaweed meal:
  Seaweed Ltd, Kylbroghlan, Moycullen, Co. Galway, Eire

Seeds, 'herbal lea' mixtures, etc.:
  Chase Organic Seeds Ltd, Gibraltar House, Shepperton, Middlesex
  Hunters of Chester Ltd, 8 Canal Street, Chester, CH1 4EJ Tel. Chester 47574
  Sinclair, McGill (Scotland) PLC, PO Box 23, 67 Kyle Street, Ayr KA7 1RY

Bibliography

All About Goats, Lois Hetherington, Farming Press, 1977
Animal Nutrition, P. MacDonald, R. A. Edwards & J. F. D. Greenhalgh, Oliver & Boyd, 2nd ed, 1973
The Book of the Goat, H. S. Holmes Pegler, Link House Publications
BRITISH GOAT SOCIETY PUBLICATIONS
   Breeds of Goats
   Dairywork
   Goatfeeding
   Goatkeeping
   Goats' Milk in the Treatment of Infantile Eczema, J. B. Tracey
   The Herd Book
   The Monthly Journal
   Therapeutic Uses of Goats Milk in Modern Medicine, Vera Walker
   The Year Book
British Poisonous Plants, H.M.S.O.
Common Ailments of the Dairy Goat, J. Laing, Scottish Goat Keeper's Association Booklet*
'The Composition and Characteristics of Goats Milk', S. Parkash & R. Jenness, Dairy Science Abstracts, 30 (2) 67–87, 1968
El Ganado Cabrio, C. Sanz Egana, Espasa Calpe, Madrid
Exhibition and Practical Goat-keeping, Joan Shields, Spur Publications, 1977

  * Enquiries to: Hon. Secretary, Scottish Goat Federation, per the British Goat Society.

The Goat Keeper's Guide, Jill Salmon, David & Charles, 1976
Goat Production in the Tropics, C. Devendra & M. Burns, Commonwealth Agricultural Bureau, 1970
Goat's Milk Diet, A Different Approach, J. Laing*
Herbal Handbook for Farm and Stable, Juliette de Baïracli Levy, Faber & Faber, 3rd ed, 1973
Home Goat Keeping, Lois Hetherington, E.P. Publishing, 1977
The Husbandry and Health of Goats, proceedings of a seminar of the Sheep Veterinary Society, 1978
La Chèvre, sous la direction de E. Quittet, La Maison Rustique, Paris
Making Cheeses, Susan Ogilvy, Batsford, 1976
Management and Disease of Dairy Goats, S.B. Gurs, Dairy Publishing Corporation, Scottsdale, Arizona, 1979
Modern Milk Goats, I. Richards, Lippincott
Nutrient Requirements for Ruminant Livestock, Commonwealth Agricultural Bureau for Agricultural Research Council, 1980
Practical Guide to Small-Scale Goatkeeping, B. Luisi, Rodale Press, 1979
(The Landsman's Bookshop, Birchenhill, Bromyard, Hereford, is a valuable source for relevant literature)

* Obtainable from: Tormoulin, Castle Douglas, Kirkcudbrightshire, Scotland DG7 3QS

Index

366   *Index*